The Reproductive & Nervous Systems

Titles in the series

The Muscular & Skeletal Systems

The Circulatory & Respiratory Systems

The Digestive & Urinary Systems

The Reproductive & Nervous Systems

THE HUMAN BODY & THE ENVIRONMENT

How our surroundings affect our health

The Reproductive & Nervous Systems

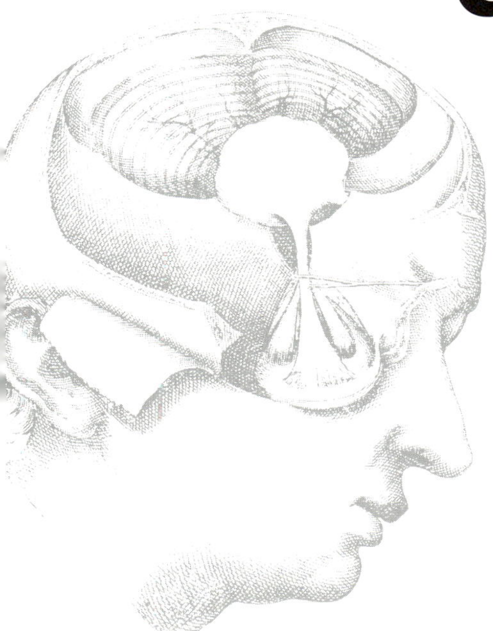

GREENWOOD PRESS
Westport, Connecticut · London

Library of Congress Cataloging-in-Publication Data

Creative Media Applications
The human body & the environment/how our surroundings affect our health
p. cm.—(Middle school reference)
Includes bibliographical references and index.
Contents: v. 1. The muscular & skeletal systems—v. 2. The circulatory & respiratory systems—v. 3. The digestive & urinary systems—v. 4. The reproductive & nervous systems.
ISBN 0-313-32558-8 (set: alk. paper)—0-313-32559-6 (v.1)—0-313-32560-X (v.2) —0-313-32561-8 (v.3)—0-313-32562-6 (v.4)
1. Human physiology—Juvenile literature. [1. Human physiology. 2. Environmental health.] I. Title: Human body and environment. II. Series.
QP37.H8925 2003
612—dc21 2002035217

British Library Cataloguing in Publication Data is available.

Library of Congress Catalog Card Number: 2002035217

ISBN: 0–313–32558–8 (set)
0–313–32559–6 (vol. 1)
0–313–32560–X (vol. 2)
0–313–32561–8 (vol. 3)
0–313–32562–6 (vol. 4)

First published in 2003

Greenwood Press, 88 Post Road West, Westport, CT 06881
An imprint of Greenwood Publishing Group, Inc.
www.greenwood.com

Printed in the United States of America

The paper used in this book complies with the Permanent Paper Standard issued by the National Information Standards Organization (Z39.48–1984).

10 9 8 7 6 5 4 3 2 1

A Creative Media Applications, Inc. Production
WRITER: Robin Doak
DESIGN AND PRODUCTION: Fabia Wargin Design, Inc.
EDITOR: Matt Levine
COPYEDITOR: Laurie Lieb
PROOFREADER: Barbara Francis
ASSOCIATED PRESS PHOTO RESEARCHER: Yvette Reyes
CONSULTANT: Michael Windelspecht

PHOTO CREDITS:
Cover: (*top*) © Robert Landau/CORBIS and (*bottom*) © Anthony Redpath/CORBIS
AP/Wide World Photographs *pages* vi, ix, 10, 14, 18, 23, 25, 27, 29, 36, 40, 43, 50, 54, 64, 73, 74, 75, 77, 78, 80, 87, 89, 92, 93, 95, 106, 108, 117, 119, 121, 125
© Rick Hall/Custom Medical Stock Photo *page* 4
© Articulate Graphics/Custom Medical Stock Photo *page* 5
© Educational Images/Custom Medical Stock Photo *page* 9, 34
© NMSB/Custom Medical Stock Photo *pages* 11, 12, 113
© B Wainwright/Custom Medical Stock Photo *page* 33
© M. English/Custom Medical Stock Photo *page* 46
© Logical Images/Custom Medical Stock Photo *pages* 59, 61
© L. Birmingham/Custom Medical Stock Photo *pages* 66, 101
© Unique Murphy/Custom Medical Stock Photo *page* 82

ILLUSTRATION CREDITS:
© Lifeart *pages* 1, 3, 31, 68, 69, 98, 100, 102

CONTENTS

TENNESSEE
771
519

INTRODUCTION

LIVING IN A RISKY ENVIRONMENT

People need sunshine, clean water, fresh air, and healthy foods to survive and thrive. There are some environmental agents, however, that can have a negative effect on our health. Some of these agents are natural. For example, exposure to large amounts of ultraviolet (UV) radiation from the sun has been shown to cause skin damage and cancer. Bacteria, viruses, and fungi that people inhale or ingest can cause disease and poor health. Other agents are put into the environment by humans. Each day, we come into contact with countless synthetic chemicals in the air that we breathe, the water that we drink, and the food that we eat.

Environmental health is defined as the human body's reaction to environmental agents, including both natural and human-made substances. In many cases, people have little or no control over the amount of exposure to these environmental agents. Some people, however, choose to do things that can cause or worsen medical problems. Smoking cigarettes and drinking alcohol, for example, can negatively affect a person's health. Lifestyle decisions made by one person can even affect those around him or her. For example, secondhand cigarette smoke has been proven to have a serious effect on the respiratory (RESS-pur-uh-tor-ee) systems of children and adults who are exposed to it.

Just as some lifestyle decisions contribute to ill health, other decisions can help keep the body as healthy as possible. Eating healthy foods and getting plenty of exercise are two important steps we can take to stay physically fit. Another way to stay healthy is to learn how environmental conditions affect us and how to minimize the damage from environmental agents.

Opposite:
Justin Gatlin (left) pulls ahead during the first heat of the men's 200-meter race at the NCAA Track and Field Championships in 2002. Gatlin, a sophomore at the University of Tennessee, can compete in his college league but is suspended from international competition because he tested positive for amphetamines at the World Junior Championships. The drug was contained in medication he took for years for attention deficit disorder.

THE ENVIRONMENT AND THE BODY

Like other body systems, the endocrine, reproductive, and nervous systems, as well as the sense organs, can all be seriously affected by the environment. Chemicals, radiation, and pollution can cause serious medical problems to these systems. These environmental agents have been linked to a number of serious health conditions, including endocrine cancers, reproductive-system cancers, and other problems. One type of pollution, noise pollution, can even lead to a permanent loss of hearing.

Bacteria, viruses, fungi, and other organisms can cause illnesses that can lead to permanent disability and even death. These illnesses can be contracted by eating contaminated food, breathing contaminated air, or coming in contact with the body fluids of someone who is infected. Sexually transmitted diseases (STDs), such as chlamydia and gonorrhea, are serious threats to reproductive health. Central nervous system conditions, such as meningitis, encephalitis, and poliomyelitis, can cause paralysis (puh-RAL-ih-siss), coma, and changes in mental functioning.

Although many environmental agents in the world can harm human health, it is important to remember that people can take many steps to stay healthy and prevent infections. This book is intended not to alarm people but to supply them with the knowledge they need to understand the risks and stay safe.

HOW TO USE THIS BOOK

Each volume in this series covers a number of body systems and the diseases—environmental and otherwise—that affect them. Each volume is organized to include the basic anatomy of each system, including its structure and function. Volume 4 includes the endocrine, reproductive, and nervous systems and the senses.

Nonenvironmental medical conditions and environmental medical conditions for the body systems are covered in separate chapters. Each disease listing includes the condition's causes, symptoms, and treatments or cures. Additional information may include diagnostic tools, statistics, and historical information.

Opposite:
Pollution from auto exhaust, factory emissions, and natural causes such as forest fires all contribute to unhealthy environmental conditions in many of the world's cities. Here, cars in Moscow drive with their lights on during the day in order to see through the heavy smog.

А171ТЕ
О816АМ

The final chapter in each volume offers tips on how to keep the body healthy. The glossary gives definitions and pronunciations of some more difficult words in the book. For a complete list of pronunciations, however, consult a medical dictionary. The index can be used to find a particular disorder easily.

Finally, the volumes are not meant to be used for self-diagnosis of medical conditions. Those who have health problems should always consult a medical professional.

Note: All metric conversions in this book are approximate.

1

THE ENDOCRINE SYSTEM

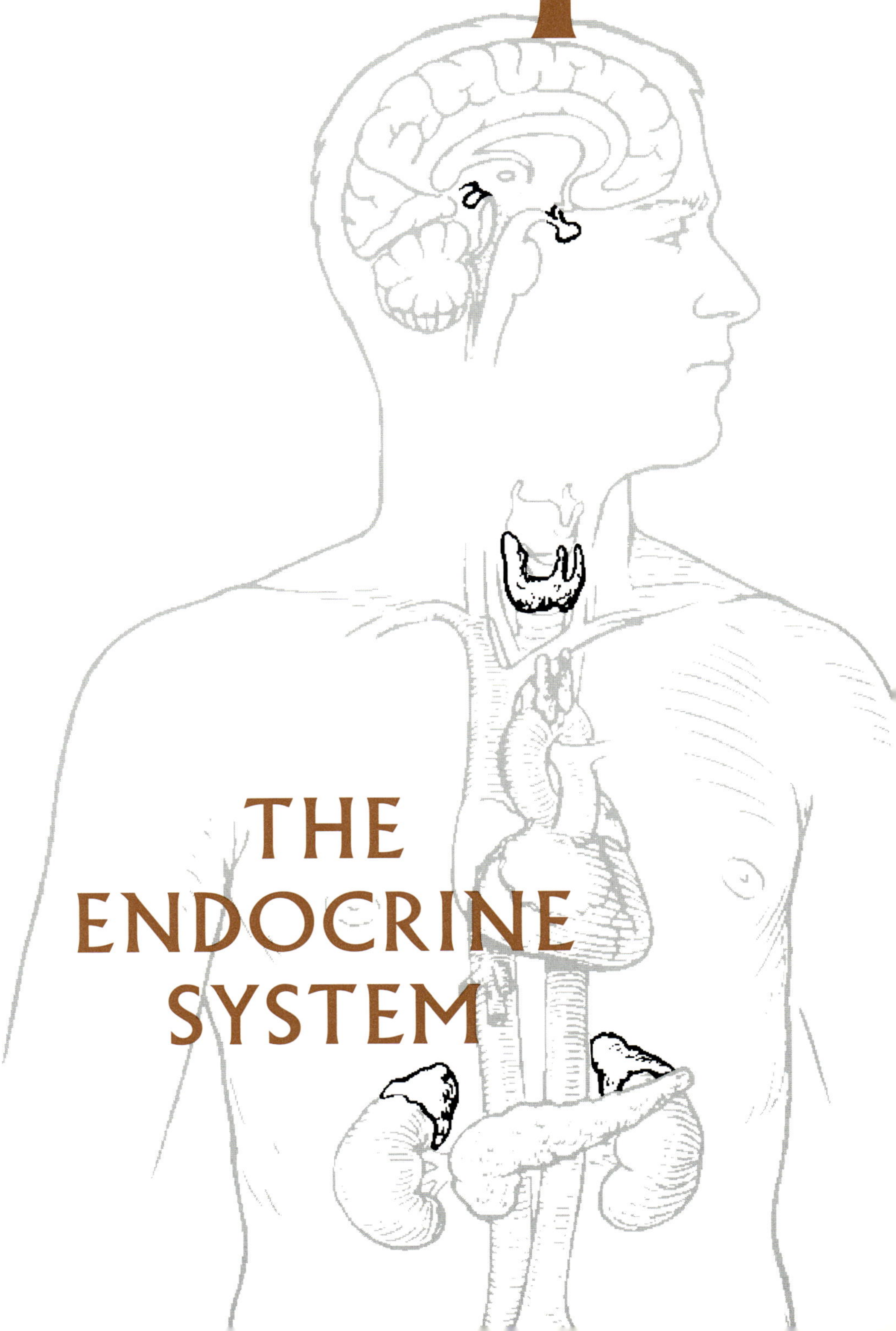

The *endocrine* (EN-doh-krin) *system* is the body system that manufactures and produces *hormones,* chemical messengers that control many of the body's functions. A hormone is a compound secreted by a gland or organ that is designed to influence the operation of a second gland or organ. Hormones regulate growth, metabolism, and other bodily functions. (*Metabolism* is the process of using energy to keep the body functioning.) The hormones' job is to keep the body working properly.

The endocrine system is made up of a network of glands. *Glands* are the groups of cells or organs that produce hormones and release them into the bloodstream. Important endocrine glands in the body include the adrenal gland, pancreas, parathyroid glands, pineal gland, pituitary gland, reproductive glands, thymus, and thyroid gland.

> **Fast Fact**
>
> Endocrine glands are not the only hormone-producing structures in the body. Certain organs, including the liver, brain, kidneys, heart, and skin, also produce hormones.

Each endocrine gland produces different types of hormones. There are more than 100 different hormones in the human body in two general classes, protein and steroid. Most of the body's hormones are *protein* hormones. These hormones are proteins that circulate throughout the body. They are broken down as needed by target organs. Protein hormones quickly stimulate body functions. *Steroid hormones,* created from cholesterol, are produced by the body as needed. (*Cholesterol* is a soft, waxy substance found in the blood and cells of the body.) These hormones usually control long-term bodily functions.

Glands secrete the hormones in response to internal or external changes in the body. These hormones travel throughout the body via the bloodstream until they reach their "target cells" or "target organs." This is the group of cells or tissues that reacts to the hormones. As the hormones travel through the body, they affect only their target cells or target organs.

Many of the body's endocrine glands are controlled by the *hypothalamus* (hye-poh-THAL-uh-muss), a section of the brain. The hypothalamus produces hormones called "releasing hormones." These releasing hormones travel to the pituitary gland, which in turn sends other activating hormones to other endocrine glands. The hypothalamus also produces hormones that help control such nervous-system functions as sleep and appetite stimulation.

GLANDS AND THEIR FUNCTIONS

The two *adrenal* (uh-DREE-nul) *glands,* located on top of the kidneys, produce hormones that help control metabolism, blood pressure, and other body functions, including nervous-system responses to stress and change. One group of hormones produced here is known as corticosteroids (kor-tih-koh-STEER-oydz). Corticosteroids perform many important functions throughout the body. One important corticosteroid is cortisol (KOR-tih-sawl), a substance that regulates blood pressure, heart and immune system functions, and the use of proteins, carbohydrates, and fats in the body. Another important hormone secreted by the adrenal glands is adrenaline (uh-DREN-uh-lin). Adrenaline, also called epinephrine (eh-pin-EF-rin), is a powerful hormone that helps the body react to stress or danger. Other hormones produced by the adrenal glands include androgens (AN-droh-jenz) and estrogen (ESS-troh-jen), hormones that are also manufactured by the reproductive glands.

The *pancreas* (PAN-kree-uss), which is also an important digestive-system organ, has special cells that produce insulin and glucagons (GLOO-kuh-gahnz). These two hormones control glucose (sugar) levels in the body.

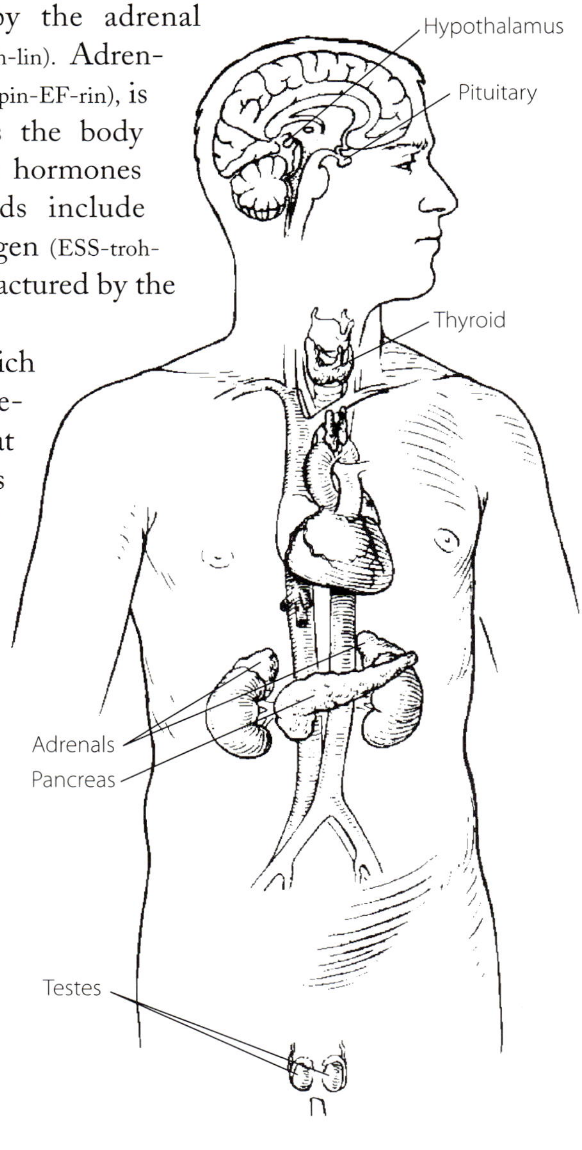

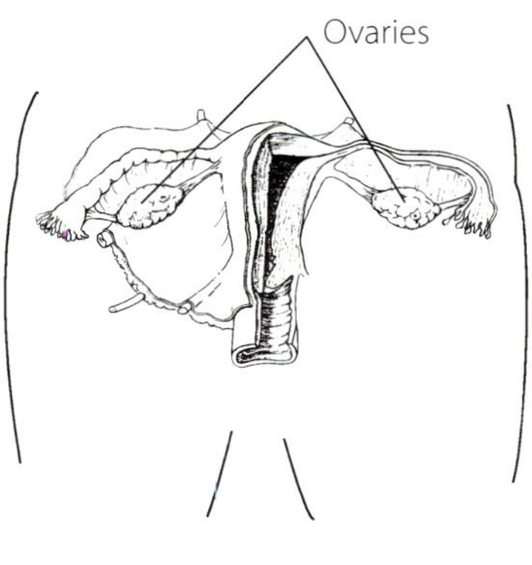

The four small *parathyroid* (paar-uh-THYE-royd) *glands,* located on the surface of the thyroid gland, produce parathyroid hormone (PTH), a substance that regulates calcium and phosphorous levels in the body.

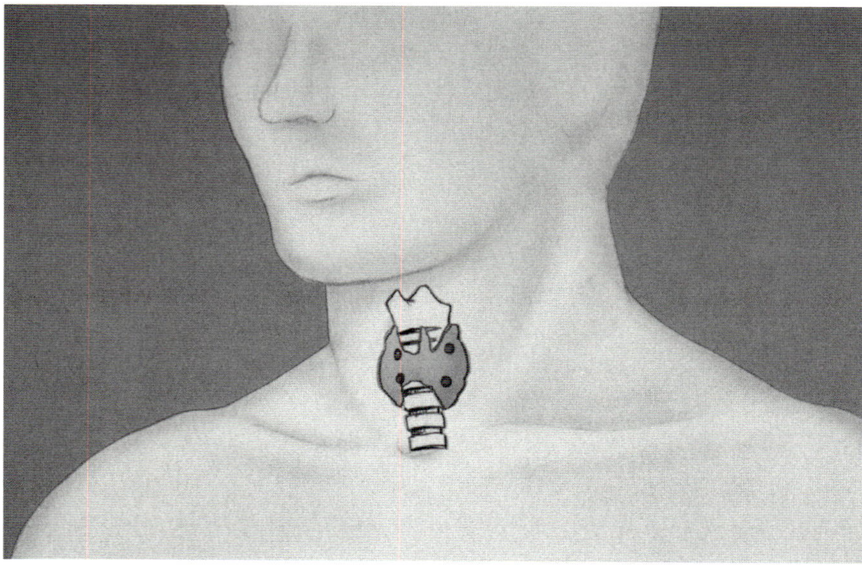

This illustration shows the location of the parathyroid glands.

> **Fast Fact**
>
> French philosopher René Descartes (1596–1650) believed that a person's soul was located inside the pineal gland.

The tiny *pineal* (PIN-ee-ul) *gland* is located deep within the brain. This gland manufactures melatonin (mel-uh-TOH-nin), a hormone that affects some daily biological functions, including sleep cycles.

The pea-sized *pituitary* (pih-TOO-ih-terr-ee) *gland* is located at the base of the brain. It is sometimes called the "master gland," because it makes hormones that stimulate hormone production and release in other glands. For example, adrenocorticotropic (uh-dree-noh-kor-tih-koh-TROH-pik) hormone (ACTH), manufactured in the pituitary gland, triggers hormone production in the adrenal gland. Another important hormone produced in the pituitary gland is growth hormone, or GH. GH regulates the growth of the body. Other hormones produced here affect kidney function and uterine-muscle function during labor.

The *reproductive glands* are those that regulate sexual development and fertility. In the female body, the main reproductive glands are the *ovaries* (OH-vuh-reez). The ovaries produce estrogen, a hormone that regulates female sexual development. In the male body, the *testes* (TESS-teez) are the main glands. The testes produce androgens, hormones that regulate male sexual development. The most important androgen in the male body is testosterone (tess-TAH-ster-ohn). Both the ovaries and the testes are important parts of the reproductive system.

Located in the upper chest, the *thymus* (THYE-muss) is believed to produce hormones that trigger certain immune-system cells to mature. After puberty, the thymus gradually disappears.

The *thyroid* (THYE-royd) *gland* is a butterfly-shaped organ found just below the larynx (LAAR-inks), where the voice box is located, in the throat. This gland produces two important hormones, triiodothyronine (trye-eye-oh-doh-THYE-roh-neen), commonly known as T3, and thyroxine (thye-ROKS-een), known as T4. These two hormones help control organ functioning and also control body heat and metabolism. Another hormone produced by the thyroid is calcitonin (kal-sih-TOH-nin), which regulates calcium levels in the body.

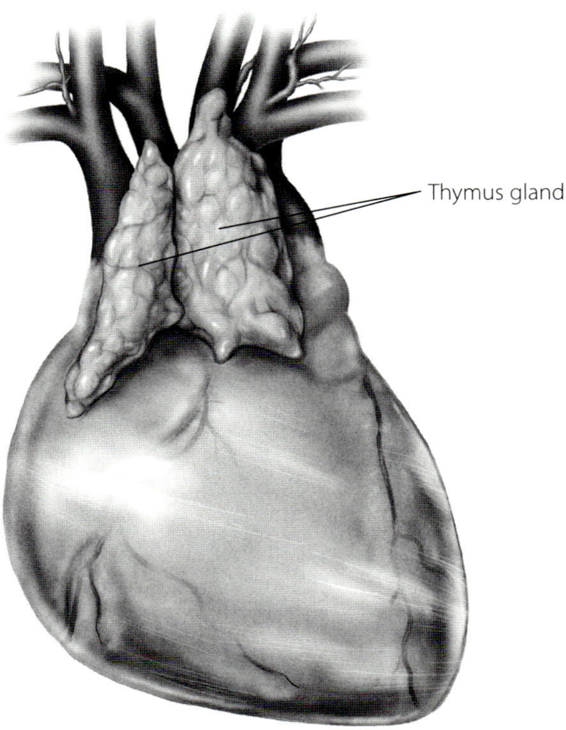

The thymus gland is located near the heart in the upper chest.

Puberty

Puberty is the physical changes that boys and girls experience as they reach sexual maturity. For boys, these changes include increased height and weight, increased amounts of facial hair, body hair, and acne; growth of the penis and testicles; and a deepening voice. For girls, puberty means increased height and weight, the development of breasts, underarm hair, and pubic hair; and the beginning of menstruation. Puberty begins when the hypothalamus secretes certain hormones that trigger the pituitary gland to release GH. In healthy children, puberty usually begins between the ages of nine and sixteen and lasts for about seven years.

2

ENDOCRINE DISORDERS

Endocrine-system disorders occur when a hormone-producing gland makes too much or too little of a hormone. When a gland produces more hormones than is normal, it is called hyperfunction. When the gland does not produce enough of a hormone, it is known as hypofunction. Even the slightest imbalance in body hormones can cause problems.

MULTIPLE ENDOCRINE NEOPLASIA

Multiple endocrine neoplasia (nee-oh-PLAY-zhuh), or MEN, is a disorder that can affect many different glands of the body. The condition involves the growth of tumors, either *benign* (harmless) or *malignant* (cancerous), on the endocrine glands. There are different types of MEN, including MEN-I (also known as Wermer's syndrome) and MEN-II (also known as Sipple's syndrome). MEN affects men and women of all ages and races.

MEN-I is an inherited disorder in which several types of endocrine glands become overactive. Usually, the first glands to be affected are the four parathyroid glands. Other affected glands most often include the thyroid and pancreas. Overactive parathyroid glands can cause an increase in the levels of calcium in the body. This, in turn, can lead to kidney stones and kidney damage. An overactive pancreas can lead to diarrhea and ulcers, while an overactive thyroid can cause increased breast milk production in women and infertility in both men and women.

Although there is no cure for MEN-I, doctors can treat the condition effectively. They usually treat parathyroid problems associated with the disease by removing nearly all of the parathyroid glands. If tumors have formed on the pancreas, doctors may prescribe strong medicine to shrink or remove the tumors. Doctors treat any thyroid tumors with medicines, radiation, or surgery, as necessary. (In radiation treatment, patients undergo doses of radiation in order to shrink tumors or cancerous areas.)

MEN-II, like MEN-I, is an inherited condition. MEN-II, however, is much more serious. With this disorder, cancerous tumors grow on the thyroid while benign tumors affect the adrenal glands. The thyroid tumors associated with MEN-II are aggressive (quick-spreading), and the condition can prove fatal if not treated quickly.

Symptoms of MEN-II include headaches, chest and abdominal pains, sweating, rapid heart rate, loss of weight, diarrhea, back pain, nausea, depression, and personality changes. Doctors treat the condition by surgically removing the thyroid, any surrounding lymph nodes, and any tumors on the adrenal glands. Afterward, the patient must undergo hormone replacement therapy to stay healthy.

ADRENAL DISORDERS

Conditions in which the adrenal glands don't produce enough of a hormone are known as *adrenal insufficiencies.* Primary adrenal insufficiency results when the adrenal glands themselves are damaged. Secondary adrenal insufficiency is caused by damage to the pituitary gland, which affects secretion of a release hormone that activates cortisol in the adrenal gland.

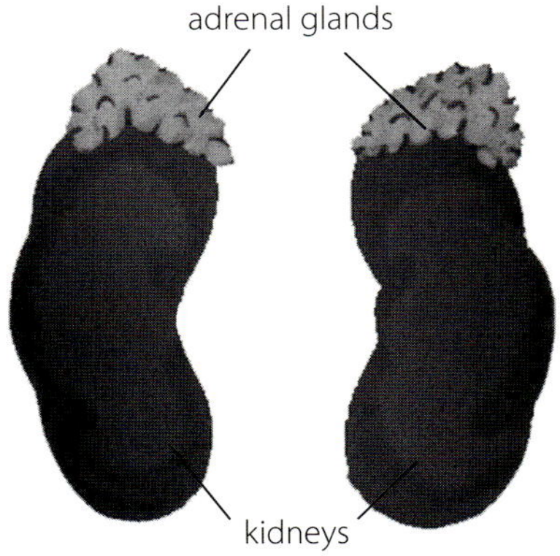

The adrenal glands are located on top of both kidneys.

ADDISON'S DISEASE

Addison's disease, a type of adrenal insufficiency, occurs when the adrenal glands don't produce enough cortisol. In some people, the adrenal glands also don't produce enough aldosterone (al-DAH-stuh-rone), a hormone that helps regulate metabolism. Seven out of ten cases of Addison's disease are autoimmune (aw-toh-im-MYOON) disorders, caused when the immune system attacks and damages the adrenal glands. This rare condition can affect young and old alike, male or female, of any race or ethnicity.

There are two types of Addison's disease, type I and type II. Type I occurs in children. Affected children may also experience parathyroid problems and other medical conditions. Type II affects young adults and may lead to diabetes, slow sexual development, and other health problems.

Symptoms of Addison's disease include weight loss, low blood pressure, muscle weakness, and darkening of the skin. These symptoms usually appear slowly. More serious symptoms, known as adrenal crisis, include pain in the lower back, abdomen, or legs, vomiting, diarrhea, low blood pressure, and fainting. If left untreated, adrenal crisis can lead to death.

Doctors treat Addison's disease with a medication that replaces cortisol in the body. Before such medicines were developed, people with this disorder almost always died prematurely as a result. People suffering from adrenal crisis may need to be hospitalized so that they can receive fluids intravenously to balance body hormones.

Health and History: JFK and Addison's Disease

John F. Kennedy (1917–1963), the thirty-fifth president of the United States, suffered from Addison's disease. The president made sure that his illness was kept top secret. He did not want the American people to worry about his health. Even years after his assassination, Kennedy's friends and family refused to allow anyone to examine his health records. Only in 2002, nearly four decades after his death, did new information reveal that Kennedy had taken several medications, including hormones, to stay healthy.

Although President John F. Kennedy, shown here playing golf at the Hyannis Port Country Club in Massachusetts, seemed to be the picture of health, he suffered from Addison's disease.

CUSHING'S SYNDROME

Cushing's syndrome, also called hypercortisolism (hye-per-KOR-tih-sawl-iz-um), is a condition that results from an overproduction of cortisol by the adrenal gland. About seven out of ten cases of

Cushing's syndrome are caused by pituitary tumors that secrete the hormone that stimulates cortisol production in the adrenal gland. Other cases are caused by tumors growing on the adrenal glands themselves. The condition most often affects women.

Symptoms of Cushing's syndrome include abnormal weight gain, especially in the upper body, thin, easily bruised skin, menstrual irregularity, bone loss, muscle weakness, and increased infection. Cushing's syndrome can lead to other medical problems, including an increased risk of heart problems, nervous-system disorders, and diabetes. The severity of the symptoms depends upon how much cortisol is being released into the body. The more cortisol that is released, the more severe the symptoms.

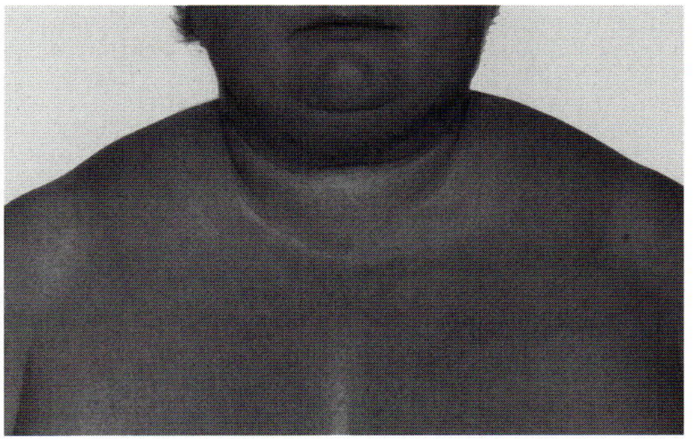

An abnormally heavy upper body is one symptom of Cushing's syndrome.

Most people with Cushing's syndrome can be successfully treated and cured of the condition. Doctors must first discover what is causing a patient's Cushing's syndrome before they can treat it. Doctors may surgically remove tumors of the pituitary and adrenal glands. Medicines and radiation therapy to shrink tumors and restore the proper cortisol levels in the body are other possible courses of treatment.

PITUITARY DISORDERS

The most common cause of pituitary-gland disorders is pituitary tumors, also known as adenomas (ad-NOH-muhz). These tumors are usually not cancerous, and they are not considered brain tumors. There are two types of pituitary tumors, secretory and nonsecretory. Secretory tumors produce excess amounts of hormones, while nonsecretory tumors take up space and interfere with the normal functioning of the pituitary gland. Some tumors may grow large enough to press on other parts of the brain, causing headaches and vision problems.

Other types of pituitary-gland disorders include the following:

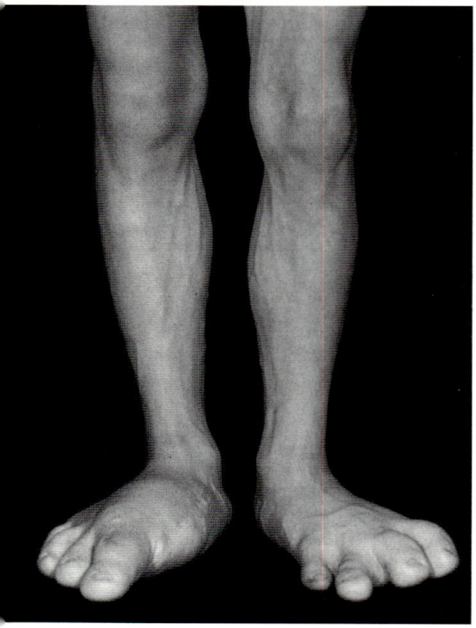

Gigantism can result in deformed legs and feet.

GIGANTISM

Gigantism (jye-GAN-tiz-um) is a condition in which children develop GH-producing tumors of the pituitary. Because the long bones are still growing, children's height is affected. The muscles and organs also grow abnormally large. In many cases, gigantism appears as part of another endocrine-related medical condition.

Symptoms of gigantism include abnormal growth during childhood, thick facial features and a protruding jaw, large hands and feet, sweating, weakness, headache, irregular menstruation, and delayed puberty. To treat gigantism, doctors may perform surgery to remove the pituitary tumor. Other treatments include radiation therapy and medications to control GH secretion.

GROWTH HORMONE DEFICIENCY

Growth hormone deficiency (GHD), also known as pituitary dwarfism, occurs when the body does not produce enough GH. Doctors are not sure exactly what causes most cases of GHD. They do know, however, that patients may also be deficient in other pituitary hormones.

The first signs of the disorder usually occur when a two- or three-year-old child's growth slows or stops prematurely. In children, other symptoms of GHD include excess fat around the face or abdomen, shortness, and slow sexual maturity. In adults, symptoms of the condition include loss of bone mass, thinning skin, weakening muscles, depression, and a lack of energy. Doctors treat GHD in children by giving patients a daily shot of GH.

PROLACTINOMA

A *prolactinoma* (proh-lak-tih-NOH-muh) is a pituitary tumor that causes excess production of the hormone prolactin. Prolactin stimulates the production of breast milk during pregnancy. Prolactinomas are the most common type of pituitary tumors:

About four out of every ten pituitary tumors are prolactinomas. Researchers are not sure what causes the growth of this type of pituitary tumor. The condition affects fourteen out of 100,000 Americans, men and women alike.

Symptoms of prolactinoma include infertility and irregular menstrual cycles in women and impotence, headaches, and eye problems in men. Doctors treat the condition with medicines to control the amount of prolactin in the body or shrink the pituitary tumor. In serious cases, surgery is often recommended.

TYPE 1 DIABETES

Insufficient production of insulin in the pancreas can cause serious medical conditions. The most common and well-known disorder related to hormone production problems in the pancreas is diabetes.

Diabetes is a disease in which a person's body cannot properly absorb sugar and starches. As a result, high levels of sugar remain in the bloodstream. There are two main types of diabetes, type 1 and type 2. Type 2 diabetes occurs when the pancreas cannot produce enough insulin.

Also known as insulin-dependent diabetes mellitus (IDDM), *type 1* diabetes is an autoimmune disorder in which the immune system attacks and kills cells in the pancreas that produce insulin. Researchers are not sure what triggers this inappropriate autoimmune response. About half of all people who develop type 1 diabetes do so before they are eighteen years old. As a result, type 1 diabetes is sometimes known as "juvenile-onset diabetes." Most children who have diabetes have this type of diabetes.

> **Fast Fact**
>
> One in 300 children is affected by type 1 diabetes.

Symptoms of type 1 diabetes include excessive thirst, hunger, and urination; weight loss; fatigue; and blurred vision. Type 1 diabetes can cause other serious health complications, including kidney, heart, and nervous-system damage. In some cases, ketoacidosis (kee-toh-ass-id-OH-siss), a diabetes-related buildup of acids in the body, causes coma and death. Treatment for this condition always includes insulin shots to replace the missing hormone. In addition, people with this disorder must maintain a healthy lifestyle through diet, exercise, and carefully monitoring blood-sugar levels.

In severe cases, doctors may perform a pancreas transplant. During this surgical procedure, doctors remove the defective pancreas and replace it with a healthy one. If a diabetic patient's kidneys have been damaged as a result of the disease, a kidney transplant may be performed at the same time. Healthy pancreases are taken from donors who have been declared brain-dead but have been kept alive by life-support machines. Kidneys, however, may be taken from living donors. Most people who receive new pancreases are under the age of fifty.

Recent advances have allowed researchers to identify certain genes that put a person at higher risk of developing diabetes. In the future, they may be able to develop medicines to prevent those at risk from ever developing the disorder. If they can identify what triggers the autoimmune response that causes diabetes, they may be able to stop people from developing the condition.

At the age of two, Jamie Langbein (left) was diagnosed with life-threatening type 1 diabetes. She has her fingertips pricked ten times a day to check her blood sugar. Her family is excited about research that may make it possible to turn embryonic stem cells into insulin-producing pancreatic cells within Jamie's lifetime.

New Technology to Treat Diabetes

In the past few years, researchers have looked for ways to improve the care and treatment of people with diabetes. One recent advance is the © Glucowatch, which in 2002 was approved for use by children and adolescents with diabetes. This device, which looks like a wristwatch, is worn around the wrist and monitors the blood-sugar levels of diabetic patients. Six times an hour, the watch takes painless blood measurements through the skin. If there is a problem with the glucose levels, an alarm goes off. Patients can then double-check their blood-sugar levels with a more traditional finger-stick test.

PARATHYROID DISORDERS

Some medical conditions affect the production of parathyroid hormones. These conditions can result in excess hormone secretion or insufficient hormone secretion.

PRIMARY HYPERPARATHYROIDISM

Primary hyperparathyroidism (hye-per-paar-uh-THYE-royd-iz-um) is a condition in which the parathyroid glands produce excessive amounts of hormone. This disorder is most often the result of a benign tumor forming on a parathyroid gland. Doctors are not sure what causes these tumors to develop. Each year, about 100,000 Americans are diagnosed with this condition. Primary hyperparathyroidism affects more women than men, especially those over the age of sixty.

Many people with primary hyperparathyroidism do not have any symptoms. Others may experience weakness, fatigue, or aches and pains. As the condition becomes more serious, a person may experience nausea and vomiting, constipation, confusion or memory loss, and increased thirst and urination. Complications from primary hyperparathyroidism include kidney stones, ulcers, high blood pressure, and problems with the pancreas.

For patients with mild cases of primary hyperparathyroidism, no treatment may be needed. Instead, doctors will monitor the condition to make sure that it is not causing other health problems. For those who need treatment, doctors surgically remove the affected parathyroid gland. This surgery cures 95 percent of all people suffering from the disorder.

HYPOPARATHYROIDISM

Hypoparathyroidism (hye-poh-paar-uh-THYE-royd-iz-um) is a condition in which the parathyroid glands do not secrete enough parathyroid hormone. As a result, calcium levels in the blood decrease, while phosphorous levels increase. The condition is most often caused by an injury to a parathyroid gland during head or neck surgery.

Symptoms of the condition include tingling in the lips, fingers, and toes; muscle cramps or spasms; facial or abdominal pain; dry hair or skin; brittle nails; and seizures or convulsions. Doctors treat the condition by prescribing calcium supplements for their patients. People suffering from hypoparathyroidism usually have to take calcium supplements for the rest of their lives.

> **Fast Fact**
>
> The prefix *hyper* comes from a Greek word meaning "over" or "excessive." The prefix *hypo* comes from a Greek word meaning "under" or "below."

REPRODUCTIVE GLAND DISORDERS

Reproductive-gland disorders most often affect sexual development and function. Reproductive-gland disorders include the following:

HYPOGONADISM

Hypogonadism (hye-poh-GOH-nad-iz-um) is a condition in which the testes or ovaries produce low amounts of hormones. *Gonads* is the medical term for the ovaries and the testes. This disorder can affect both males and females of any age. Males have a testosterone deficiency while females experience an estrogen deficiency. Primary hypogonadism is a problem in the testes or ovaries. Secondary hypogonadism is a problem in the pituitary gland, which secretes the release hormones for testosterone and estrogen.

Hypogonadism can be caused by a number of factors, including genetic defects, chemotherapy and radiation treatments, testicular injury, and some types of surgery. Aging also typically decreases the amount of testosterone produced by the body.

Hypogonadism can affect the body in different ways, depending upon when the condition occurs. If hypogonadism occurs during fetal development, for example, then a child may be born with undeveloped sexual organs. This is called *ambiguous genitalia* (am-BIG-yoo-us jen-ih-TAYL-ya). When the condition occurs at puberty, it affects the normal development of sexual characteristics. Girls, for example, may not menstruate or develop breasts, while boys may experience growth problems. When adult males develop the condition, it can cause erectile dysfunction, infertility, an increase in body fat, loss of bone mass, and other problems. Adult females may stop menstruating and suffer hot flashes and loss of hair on the body.

Doctors treat some cases of hypogonadism with hormone replacement therapy. If the condition is caused by a pituitary problem, doctors can treat hypogonadism with pituitary hormones or by surgically removing any pituitary tumors.

POLYCYSTIC OVARY SYNDROME

Polycystic (pahl-ee-SISS-tik) *ovary syndrome* (PCOS), also known as Stein-Leventhal syndrome, is believed to be caused by an overproduction of androgens in the ovaries. An excess of androgens causes eggs inside the ovaries to develop into small *cysts,* or fluid-filled sacs. The ovaries then become enlarged and develop a tough, white outer layer. Most women who develop PCOS do so shortly after puberty.

A main symptom of PCOS is irregular or absent periods as a result of eggs failing to leave the ovaries. Other symptoms include cramps, bloating, weight gain and obesity, acne, excess facial and body hair, infertility, and high blood pressure. Complications of the disorder include diabetes, heart disease, uterine bleeding, and cancer.

Doctors use medications to treat PCOS and its symptoms. Birth control pills, for example, may be used to clear up acne and stop excess hair growth. Other medicines that stimulate ovulation may also be used. *Ovulation* is the release of a mature egg from the ovaries.

Health in the News: Hormone Replacement Therapy

One of the standard treatments for women going through menopause is hormone-therapy treatment, particularly estrogen and progesterone (proh-JESS-tuh-rone) replacement. According to the National Women's Health Information Center, about 40 percent of all postmenopausal women are receiving hormone replacement therapy. The benefits of this therapy are well documented. Hormone replacement therapy decreases the risks of osteoporosis (ah-stee-oh-puh-ROH-siss)—a condition that results in brittle, fragile bones—for postmenopausal women. The treatment may also cut the chances of getting colon cancer and developing type 2 diabetes. It also relieves certain symptoms of menopause, including hot flashes, night sweats, vaginal dryness, and bladder infections.

However, there are also risks involved in this treatment. It has long been known that estrogen raises the risk of some types of uterine and ovarian cancers. Taking progestin (synthetic progesterone) with estrogen was meant to lessen that risk. Now, it seems that the two hormones, taken together, increase the odds of some women developing breast cancer. In addition, new studies have shown that hormone therapy does not reduce the risk of heart disease, as previously thought. In fact, taking the two drugs together may increase the risk of stroke, heart disease, and blood clots.

According to researchers at the National Heart, Lung, and Blood Institute (NHLBI), about 60 percent of women face greater risks than benefits from hormone-replacement therapy. However, most doctors and medical experts agree that more study is needed. They believe that hormone therapy is still an important and effective means of keeping postmenopausal women healthy.

Jean Golub holds a packet of hormone-replacement therapy drugs in the living room of her home in Shaker Heights, Ohio. Golub is one of many women who are reconsidering the pros and cons of hormone replacement therapy in light of new research findings.

THYROID DISORDERS

Disorders of the thyroid gland are the most common endocrine-system disorders. About 12 million Americans are affected by thyroid conditions. Women are more likely than men to suffer from these conditions.

The thyroid can be affected in a number of ways. The most common type of thyroid condition is *hypothyroid* (hye-poh-THYE-royd), or underactive thyroid. However, *hyperthyroidism* (hye-per-THYE-roid-iz-um) is another condition, and the thyroid can suffer from irritation, inflammation, and tumors, as well. Thyroid-gland disorders include the following:

GRAVE'S DISEASE

Grave's disease is an autoimmune disorder that occurs when the body's immune system mistakenly attacks the thyroid, causing it to produce an excess of thyroid hormones. Also known as overactive thyroid, Grave's disease is the most common type of hyperthyroidism. The condition most often affects women over the age of twenty, but anyone may develop the disorder. About 2 percent of all women in the United States suffer from Grave's disease.

Symptoms of Grave's disease include nervousness, heart palpitations, weight loss, insomnia, sweating, prematurely gray hair, muscle weakness, and sensitivity to cold temperatures. The condition may also result in protrusion of the eyes, a condition known as exophthalmos (eks-ahf-THAL-moss). Exophthalmos can cause tearing and irritation of the eyes. Although there is no current cure for Grave's disease, it can be treated effectively. Doctors use medicines, radiation, or surgery to control excessive thyroid activity.

HASHIMOTO'S THYROIDITIS

One of the most common types of hypothyroidism is *Hashimoto's thyroiditis* (hah-shee-moh-toze thye-roy-DYE-tiss). Hashimoto's thyroiditis, also called chronic thyroiditis, is an autoimmune disorder in which the immune system attacks and damages the thyroid gland. This type of hypothyroidism affects about 5 million Americans, most of them middle-aged women.

Symptoms of Hashimoto's thyroiditis include fatigue, loss of energy, depression, weight gain, dry skin, hair loss, constipation, infertility, and irregular menstrual cycles in women. The condition can also lead to a slowing of heart and lung functioning. Severe hypothyroidism, called myxedema (miks-eh-DEE-muh), can lead to coma and death.

The treatment for hypothyroidism is hormone replacement therapy. Patients are most commonly given T4, but they may also receive T3 or a combination of the two thyroid hormones. Some people have to undergo this treatment for the rest of their lives.

DIAGNOSING ENDOCRINE DISORDERS

Doctors use a number of different tests to detect and diagnose endocrine problems. Two important tests are blood and urine tests. Doctors look at hormone levels in the blood and urine in order to learn which endocrine glands are causing a problem. Other tests include *magnetic resonance imaging* (MRI) tests or *computerized tomography* (tuh-MAH-gruh-fee) scans (CT scans). These are imaging tests that allow doctors to look into the body for tumors that may be causing endocrine disorders.

Other more specialized tests can also be used to find glandular problems. For example, doctors may inject insulin into a patient's body in order to observe how various parts of the body react. During a *thyroid scan,* patients drink radioactive iodine so that doctors can check for thyroid problems. Doctors may also use biopsies (BYE-ahp-seez) to take cell samples from suspect glands. A *biopsy* is a surgical procedure to remove tissue, cells, or fluids from the body for examination.

Doctors use a number of treatments to help patients suffering from endocrine-system disorders. *Radioactive iodine therapy* is sometimes used to destroy thyroid cancer cells. Patients take radioactive iodine—also called I-131—by mouth. The iodine is in the form of either a liquid or a capsule. Another effective treatment for some glandular disorders is hormone-replacement therapy. With this type of treatment, patients are given synthetic hormones that replace the ones they are lacking.

3

HOW THE ENVIRONMENT AFFECTS THE ENDOCRINE SYSTEM

The environment can have a serious effect on the endocrine system. Some chemicals and drugs, for example, have been proven to disrupt the body's hormonal balance. Other environmental agents, such as smoking, alcohol use, and poor nutrition, can also cause endocrine-system problems.

CHEMICALS, DRUGS, AND RADIATION

Endocrine disruptors are chemicals, whether synthetic or naturally occurring, that interfere with the proper functioning of the endocrine system. Endocrine disruptors can be found in the air that we breathe, the water that we drink, and the food that we eat. Children exposed to these chemicals are at a higher risk of being harmed than adults are. That's because their bodies are still growing.

According to the Environmental Protection Agency (EPA), a government group that works to protect human health and the natural environment, endocrine disruptors can affect the endocrine system in a number of ways. Some mimic, or imitate, hormones, tricking the body into a response that it shouldn't be making. Chemicals called environmental estrogens, for example, mimic the hormone estrogen. Other chemicals can block hormones. Still others stimulate or stop hormone production in certain glands.

The EPA has the power to ban or strictly regulate any chemicals that are shown to harm the endocrine system. Some endocrine disruptors have already been banned or regulated. One chemical that was shown to be especially harmful was diethylstilbestrol (dye-eth-ul-stil-BESS-trawl), or DES. DES is an artificial estrogen that was used between 1940 and 1971 to prevent pregnant women from miscarrying. After researchers proved that DES caused birth defects, especially of the fetal reproductive system, the chemical was banned. Women who used DES may have a higher risk of breast cancer. The drug was also shown to increase the risk of vaginal and cervical cancer in the daughters of women who took it.

Other endocrine disruptors include a group of chemicals known as organochlorine (or-gan-oh-KLOR-een) and polychlorinated compounds. These compounds include polychlorinated biphenyls (PCBs), which were used as coolants, and the pesticides dichlorodiphenyltrichloroethane (dye-klor-oh-dye-fee-nul-trye-klor-oh-ETH-ayn), or DDT, and dioxin (dye-OKS-in). Dioxin was one of the chemicals used to make Agent Orange, an herbicide used by the United States military in South Vietnam during the Vietnam War (1964–1975). *Herbicides* are chemicals used to destroy plants. After the war, Vietnam veterans began having numerous health problems, including increased incidences of the cancers Hodgkin's disease, non-Hodgkin's lymphoma (lim-FOH-muh), multiple myeloma (mye-uh-LOH-muh), and the skin condition chloracne (klor-AK-nee).

The National Football League has taken a strong position against the use of anabolic steroids by its players. Players who violate the league's policy risk suspension or expulsion from the league.

HORMONE ABUSE

Anabolic steroids are synthetic hormones that are similar to androgens. They are often prescribed by doctors to treat anemia and other medical conditions. Anabolic steroids also have the ability to enhance athletic performance, speed recovery time from injuries, and add to muscle mass. As a result, some people choose to use anabolic steroids inappropriately. They take these drugs illegally as a quick way to "pump up," or make their bodies stronger and faster. Recently, more young people have begun abusing anabolic steroids to achieve these effects quickly.

What young athletes may not understand is that anabolic steroids have some serious side effects, including facial and hair changes, damage to the sexual organs, acne, dizziness, nausea, vomiting, and high blood pressure. Men who take the drugs may experience infertility and the development of breasts. Women who take anabolic steroids may stop menstruating and develop masculine characteristics, including facial and body hair.

Another side effect of these drugs is known as "steroid rage." Steroid rage is the result of wild mood swings and depression brought on by the drugs. People experiencing steroid rage may be violent or quick to anger and may make poor decisions. They may also be paranoid and suffer from delusions.

The long-term effects of anabolic steroids are still being researched. Doctors already know that steroid abuse can lead to an increased risk of heart disease, liver disease, and liver, kidney, and prostate cancer. In addition, when steroids are injected, sharing needles raises the risks of developing such serious diseases as the liver condition hepatitis (hep-uh-TYE-tiss) and the immune system condition known as *acquired immune deficiency syndrome* (AIDS).

As the dangers of anabolic steroids have become clearer, more and more people have begun speaking out against these drugs. Many sports associations have now banned the use of anabolic steroids by their athletes. These associations include the International Olympic Committee (IOC), the National Football League (NFL), and the National Basketball Association (NBA). Currently, Major League Baseball (MLB) and the National Hockey League (NHL) do not ban these drugs.

RADIATION AND THYROID CANCER

One common endocrine condition is *thyroid nodules.* Thyroid nodules are swellings of the thyroid glands. About half of all people experience these swellings at some time in their lives. Although 95 percent of all thyroid nodules are benign, a small percentage turn out to be cancerous. For this reason, it is important for anyone who notices a bump or swelling in the thyroid area to seek medical attention.

Some people are at higher risk for thyroid cancer than others are. For example, people who undergo radiation therapy for other types of cancers have a greater chance of developing thyroid cancer. White women over the age of forty are also more likely to get the condition than other populations are.

Studies have shown that people, especially children, who are exposed to radioactive fallout also have a higher incidence of thyroid cancer. In April 1986, an explosion ripped apart the Chernobyl nuclear power plant in Ukraine (then part of the Soviet Union). Radioactive materials were released into the air. Millions of people in surrounding countries were exposed to the radioactive fallout. Studies conducted years later showed that about 3,000 people who were exposed to the radiation later developed thyroid cancer. Because thyroid cancer can take between eight and twenty years to develop, the United Nations believes that many more of those exposed to the radiation may eventually develop thyroid cancer as a result of the accident.

The United States is home to a number of nuclear power plants. The Food and Drug Administration (FDA) has recommended that people who live within 50 miles (80 kilometers) of a nuclear power plant keep a supply of potassium iodide pills. These pills block the thyroid gland from absorbing radioactive iodine. After the Chernobyl disaster, Poland was quick to pass out potassium iodide pills. Fewer people there developed thyroid cancer compared with other nations that received similar exposure to the fallout. Potassium iodide pills should be taken within three hours of exposure to nuclear fallout.

Observers watch an atomic nuclear blast on March 23, 1955. The National Cancer Institute has announced that fallout from 1950s nuclear-bomb tests exposed millions of children across the country to radioactive iodine, raising the possibility that 10,000 to 75,000 of them might develop thyroid cancer. But government doctors emphasized that they have no proof that this radioactive substance causes thyroid cancer, so their estimate is a worst-case scenario.

THYROID CANCER IN THE UNITED STATES

Each year, about 20,000 Americans are diagnosed with thyroid cancer. People most at risk are those who had radiation therapy to the neck, especially when they were children. In the 1950s, radiation therapy to the neck was used to treat enlarged thyroids, adenoids (masses of lymphatic tissue in the upper throat), and tonsils (masses of lymphatic tissue on the sides of the throat), as well as some skin disorders.

There are several different types of thyroid cancer. They are classified by the types of cells in which the cancer starts. The most common type of thyroid cancer is *papillary carcinoma* (PAP-ih-leh-ree kar-sih-NOH-muh), which begins in the papillary cells of the thyroid. This type of thyroid cancer spreads slowly and is the most treatable form of the disease. As many as 75 percent of all people diagnosed with thyroid cancer have papillary carcinoma.

Symptoms of thyroid cancer include an enlarged thyroid, hoarseness, a cough (sometimes with blood), and difficulty swallowing. To diagnose the disease, doctors may use imaging tests, biopsies, and other tests that detect cancerous tumors or cells.

The chances for effectively treating and curing thyroid cancer are excellent. Before deciding upon a course of treatment, doctors must determine the type of tumor involved, as well as its stage, or level of growth and development. Like other cancers, thyroid cancers can spread from their original site to nearby lymph nodes. From here, the cancers may spread to other organs throughout the body.

In most cases, thyroid cancer can be managed through surgery to remove part or all of the thyroid gland. Other treatments that may be used in addition to or instead of surgery include radioactive-iodine therapy, hormone treatment, radiation treatment, and chemotherapy.

DIET AND LIFESTYLE

Poor nutrition can cause the endocrine system to malfunction. A lack of iodine in the diet, for example, can lead to problems with the thyroid. Goiter and thyroid cancer are two endocrine disorders that may result from iodine deficiency. *Goiter* is an enlargement of the thyroid gland. In past centuries, goiter was almost always caused by a lack of iodine in the diet. Today, however, there are a number of other factors known to cause the condition, including Grave's disease, Hashimoto's disease, thyroid nodules, and thyroid cancer. The risk of thyroid cancer has also been known to increase in those with iodine deficiency. According to the National Cancer Institute (NCI), thyroid cancer seems to be less common in the United States than it is in countries where iodine deficiency is a problem.

A Morton Salt factory worker monitors a conveyor belt of familiar salt cartons heading to a final packaging area. The Morton Salt Company was the first company to add iodine to salt to reduce the instance of goiter in children.

Health and History: Salt and Iodine

In the early twentieth century, goiter was a common problem for children in the United States. In 1924, the Morton Salt Company became the first company to add iodine to its salt in order to reduce the number of goiter cases. State health officials encouraged people to use iodized salt, and the incidence of goiter fell rapidly. In Ohio, for example, a study conducted between 1924 and 1936 showed that 31 percent of all school-age children had goiter before iodized salt was available. By 1936, the incidence of goiter was just 7 percent in children who used iodized salt.

Today, iodine is added to other products, including some types of bread. In areas where iodine is not added to food products, iodine deficiency disorders are still common. According to the National Center for Health Statistics, iodine deficiency is a major deficiency disease around the world.

SECONDARY HYPERPARATHYROIDISM

Hyperparathyroidism is the excess production of hormones by the parathyroid gland. When this condition is the result of a lack of calcium in the body, the condition is more specifically known as *secondary hyperparathyroidism.* People at risk from this condition include children who have calcium-poor diets and elderly people who do not get enough sunlight. These two populations are more likely to have calcium deficiencies in the body. Chronic kidney failure is another major cause of secondary hyperparathyroidism. This condition is most often treated with calcium supplements.

TYPE 2 DIABETES

Type 2 diabetes, also known as non-insulin-dependent diabetes mellitus (NIDDM), usually occurs in later adulthood. With this type of diabetes, the body cannot react correctly to insulin. (Type 1 diabetes is an autoimmune disorder in which the immune system attacks and kills cells in the pancreas that produce insulin.) Medical researchers have tied type 2 diabetes to obesity and lack of exercise. The condition once affected mostly overweight people over the age of forty. For this reason, type 2 diabetes is sometimes referred to as "adult-onset diabetes." However, more and more young people have begun suffering from this type of diabetes. Researchers believe that this is a result of more children being overweight and inactive. In addition, certain ethnic groups, including African Americans, Native Americans, and Latinos, are at a higher risk of developing type 2 diabetes.

Symptoms of the condition include excessive thirst, urination, and hunger, weight loss, and nausea. Secondary symptoms may include blurred vision, frequent infections, and slow healing of wounds. Type 2 diabetes can lead to serious health complications, including heart attack, stroke, kidney failure, blindness, and death.

More than 7 million Americans suffer from type 2 diabetes. Some may be able to treat the condition with diet, getting their weight under control, and eating healthy foods. Exercise can also help. For those who cannot be helped by lifestyle changes, doctors may prescribe oral medications or insulin injections.

HYPOGLYCEMIA

Hypoglycemia (hye-poh-glye-SEE-mee-uh) is a condition in which the levels of glucose, or sugar, in the blood become low. This can happen when the body uses glucose too rapidly and it is not replaced, or when the body releases too much insulin into the bloodstream. Insulin speeds up the use of glucose in the body. Hypoglycemia often results from skipping meals or poor dieting practices. However, the condition can also be caused by tumors of the pancreas, liver disease, and alcohol consumption.

Symptoms of hypoglycemia include fatigue, nausea, nervousness, shaking, sweating, headaches, blurred vision, and convulsions. People who are not diabetic can usually cure hypoglycemia quickly by eating something. Sugars and carbohydrates are especially effective for quick recovery from any hypoglycemia symptoms.

For people who are diabetic, however, hypoglycemia can cause serious medical complications, including coma or brain damage. This is called "insulin shock." Diabetics must take special care to avoid hypoglycemia. In some cases, they may need to take medications to help them control and prevent the condition.

SEASONAL AFFECTIVE DISORDER

Seasonal affective disorder (SAD) is a mental-health condition that some researchers believe might be linked to hormone production in the pineal gland. This gland produces melatonin, a hormone that affects sleep cycles. Some believe that melatonin levels rise if there is an absence of light. As a result, melatonin levels are higher in the autumn and winter, when daylight hours are short, than in the spring and summer, when daylight hours are longer. Elevated melatonin levels can cause feelings of sleepiness and depression.

Ken Hess, who suffers from seasonal affective disorder, reads in front of a special light designed to simulate sunlight at his home in Falmouth, Maine. Decreased hours of sunlight in the winter months are believed to cause symptoms of depression in about 15 percent of the population.

SAD affects women more often than men. It may begin in adolescence or early adulthood. Symptoms of the disorder include depression, fatigue, an inability to concentrate, and a craving for carbohydrates. People diagnosed with SAD can ease their symptoms by spending time each day under a special light.

4

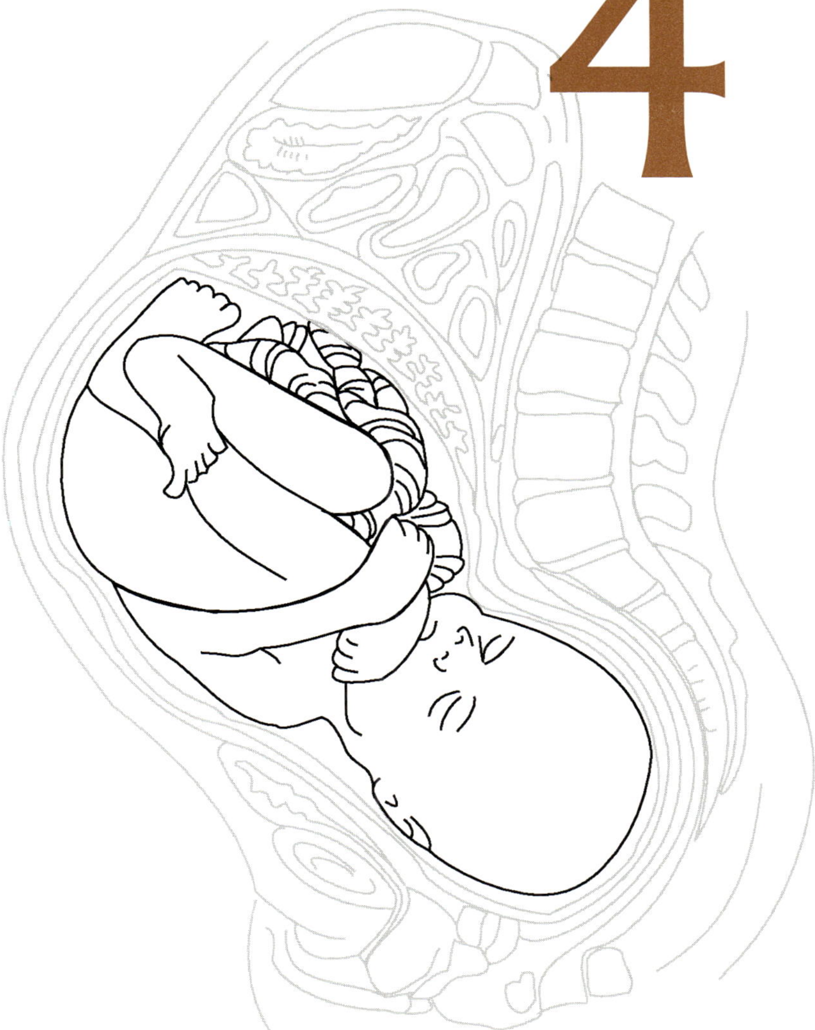

THE REPRODUCTIVE SYSTEM

The *reproductive* (ree-proh-DUK-tiv) *system* is made up of the organs and other body parts that are responsible for creating new life. The female and male reproductive systems are very different from each another. Each plays a unique role in producing a child. However, both the male and female systems must work together to create the child. The female reproductive system produces the egg, while the male reproductive system produces sperm to fertilize the egg. When the egg and the sperm meet in the female's reproductive system, new life begins.

THE FEMALE REPRODUCTIVE SYSTEM

The female reproductive system is made up of the parts of the body that are responsible for conception, pregnancy, and childbirth. The parts of the female reproductive system are often described as the external organs and the internal organs. The external organs are those that are visible on the outside of the body, while the internal organs are inside the body.

The external reproductive organs include the mons veneris (mahnz VEN-er-iss), the labia, the clitoris, and the vestibule. Together, these organs are known as the vulva. The job of the external organs is to provide a protective covering for the internal organs. These outer organs are also stimulated during sexual contact.

The internal reproductive organs include the ovaries, the fallopian tubes, the uterus, the cervix, and the vagina. Reproduction begins with the ovaries, two oval glands that produce eggs. The ovaries also manufacture the hormones estrogen and progesterone. Each month, an egg inside one of the ovaries matures and is released. This process is called *ovulation.*

The *fallopian* (fuh-LOH-pee-an) *tubes* are two muscle-lined passages that connect the ovaries to the uterus. Mature eggs pass out of the ovaries and into the fallopian tubes. An egg spends about two days in the fallopian tubes. Then, whether it has been fertilized by a sperm or not, it is expelled into the uterus.

Fast Fact

When a female baby is born, her ovaries already contain all of the approximately 10,000 egg cells that she will ever have. The egg cells do not mature until she goes through puberty.

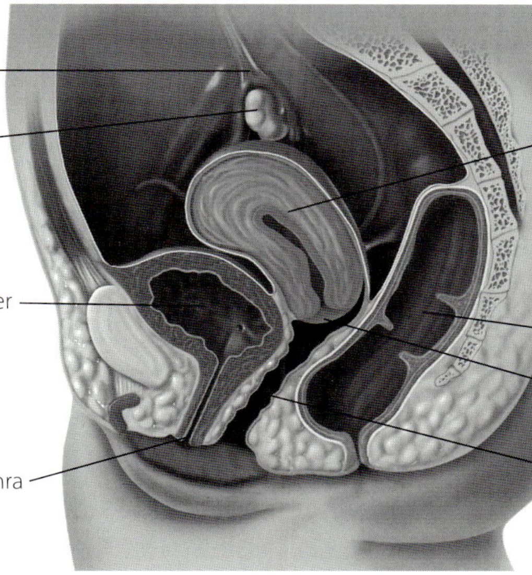

This illustration shows the parts of the female reproductive system.

The *uterus* (YOO-ter-uss) is a hollow, muscular chamber. Each month, the lining of the uterus thickens, creating a nourishing environment for a fertilized egg to attach to and grow in. It is inside the uterus that the fertilized egg will, over a period of nine months, grow and develop until it is ready to be born.

If an egg is not fertilized in the fallopian tubes, the uterus sheds its lining and expels it, as well as the egg. The lining and egg are pushed through the cervix, which is the narrow end of the uterus that connects it to the vagina. The *vagina* is a muscular passage that extends to the outside of the body.

The process of an egg maturing and being shed from the body is called the *menstrual cycle.* The menstrual cycle, which averages twenty-eight days in length, usually begins when a girl is between the ages of eight and thirteen. The cycle is triggered by reproductive hormones. The cycle continues each month until hormone production slows, usually when a woman is between forty and fifty-five. Eventually, the woman ceases having a monthly cycle. This is known as *menopause.*

THE BREASTS

The *breasts* are not technically part of the reproductive system. Mammary glands inside the female breasts, however, produce milk for an infant after childbirth. For this reason, breasts are included in this section. Both men and women have breasts.

Each breast is made up of mainly fatty tissue. This tissue contains between fifteen and twenty lobes of glandular tissue. Inside the lobes, small lobules produce milk after childbirth. Ducts connect the lobules to the *nipple,* a projection at the end of the breast. The size of a woman's breasts is determined by the amount of fatty tissue that is present.

As with other parts of the reproductive system, the breasts can be affected by aging. They begin to lose fat and tissue, becoming less firm and dense. The number of mammary ducts within the breast also decreases.

THE MALE REPRODUCTIVE SYSTEM

The job of the male reproductive system is to produce and deliver sperm to the female for fertilizing an egg. The male reproductive system is made up of the testes (also called the testicles), the epididymis, the vas deferens, and the penis. Sperm cells are produced in the *testes,* two glands that, as part of the endocrine system, also manufacture important sexual hormones. The testes are located in a protective sack outside the abdomen called the scrotum. Each day, the testes produce about 500 million sperm cells.

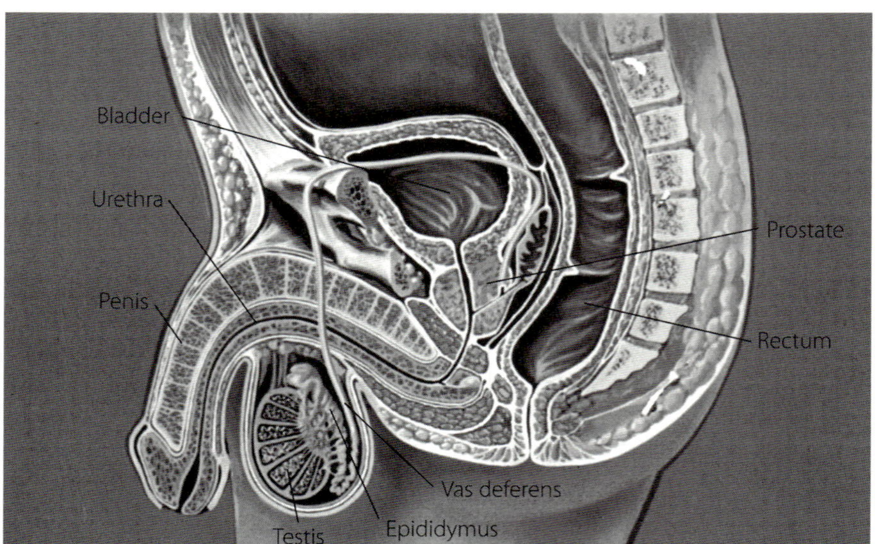

The male reproductive system is illustrated in this drawing.

After sperm cells are created, they move into the *epididymis* (eh-pih-DID-ih-miss), coiled tubes behind the testes. Here, the sperm cells mature for about two weeks. Then they travel through the *vas deferens* (vass DEF-er-enz), a tubelike structure that connects the epididymis to the seminal vesicle. The seminal vesicle is a gland that releases a fluid that helps make up semen. *Semen* mixes with sperm, and this fluid is expelled during ejaculation. The prostate gland, located near the seminal vesicle, also produces fluids that make up semen.

During ejaculation, semen travels through the ejaculatory duct to the *urethra* (yoor-EE-thruh), a tube that stretches through the entire length of the penis. The *penis* is the tube-shaped external organ of the reproductive system. During ejaculation, a male releases millions of sperm cells. Only one, however, is needed to fertilize an egg and create a new life.

Like women, men also experience reproductive-system changes as they become older. Inside the testicles, the amount of tissue decreases, as does the rate of sperm production. Other parts of the reproductive system are also affected by aging. However, men continue to produce sperm throughout their lives, so they can father children even in old age.

CONCEPTION AND BIRTH

A woman becomes pregnant when the mature egg in her fallopian tube is fertilized by a sperm. This is called *conception.* After conception, the fertilized egg, now called a *zygote,* moves into the uterus. The zygote is a single cell that contains the genetic information from both the mother and the father. In the uterus, the zygote attaches itself to the uterine wall and begins a process of growth and development that will continue for nine months.

In the uterus, the single-celled zygote begins dividing and multiplying. By the eighth week of pregnancy, it has become a *fetus,* identifiable as male or female. The fetus has developed the tissue that will eventually become its brain, heart, eyes, and other organs. By the thirty-sixth week of development, the baby can survive outside its mother's body.

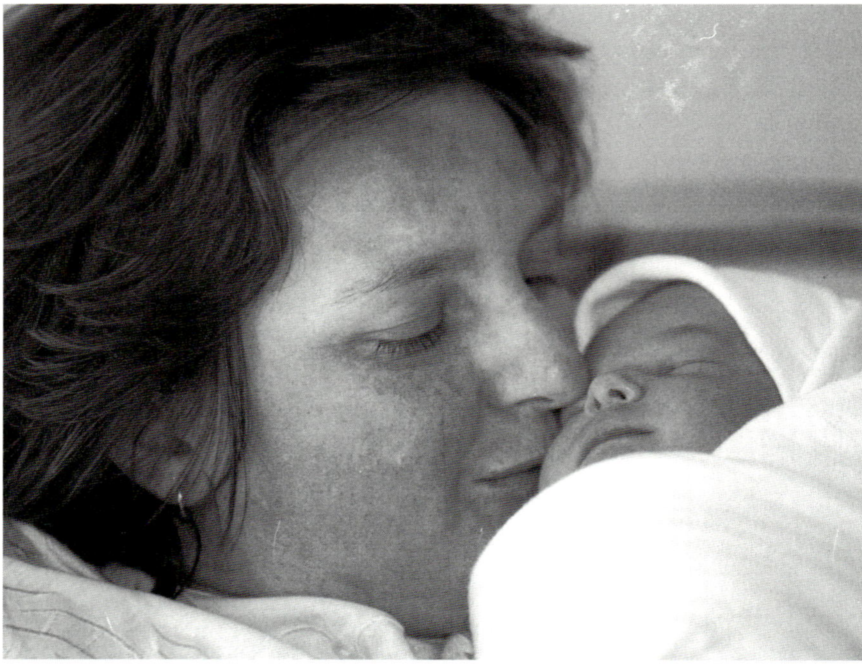

Fatima Nevic kisses her baby boy, whose birth at two minutes after midnight on Tuesday, October 12, 1999, made him the designated six billionth person on the planet.

Women usually go into labor about forty weeks after conception. Before this happens, however, a woman's body prepares to expel the infant. The uterus begins to contract, and the cervix softens and become thinner. The cervix also begins to *dilate,* or expand. This will allow the infant to squeeze through the small opening during delivery. As these changes are taking place, the infant's body rotates until its head is pointing downward. As true labor begins, uterine contractions become stronger and more painful. These contractions push the baby through the cervix and the birth canal and, finally, out into the world.

5

REPRODUCTIVE DISORDERS

Both the male and female reproductive systems can be damaged by disorders and disease. Reproductive system disorders affect people of all ages and races. They may occur at any time in a person's life.

GENETIC DISORDERS

Some reproductive disorders are inherited or the result of a birth defect. Genetic reproductive conditions include the following:

KLINEFELTER'S SYNDROME

Klinefelter's syndrome is a genetic disorder that affects only males. Male babies with this condition are born with two X chromosomes instead of the single one that is normally present in healthy males. Klinefelter's syndrome causes hypogonadism. Researchers believe this disorder is not inherited, occurring instead by chance.

Symptoms of Klinefelter's syndrome include a small penis and testicles, infertility, enlarged breasts, long arms and legs, and learning disabilities. There is no way to treat infertility caused by the disorder. Other symptoms, however, may be treated with testosterone therapy.

TURNER'S SYNDROME

Turner's syndrome is a genetic disorder in which a female is born missing one of the two X chromosomes normally present in healthy females. This syndrome is one of the most common chromosomal disorders. Females affected by this condition may suffer from infertility and abnormal sexual development.

There is no known cause for Turner's syndrome. Symptoms of the condition include shortness, abnormal eye features and bone development, absent secondary sexual characteristics (such as breast or pubic hair development), failure to menstruate, and infertility. Complications of the disorder include heart, kidney, and thyroid disorders, high blood pressure, diabetes, arthritis (arth-RYE-tiss), and obesity. Doctors sometimes treat the condition with hormones that stimulate growth and sexual development.

PREGNANCY AND LABOR PROBLEMS

There are a number of health conditions that can affect a man or woman's ability to produce a child. An inability to produce children is known as *infertility.* Once a woman does become pregnant, there are a number of medical conditions that can affect her or her fetus. Some of these conditions may be brought on by environmental agents. Others are a result of genetic defects. In addition, problems can occur during labor that put both mother and child at risk. Infertility and problems with pregnancy and labor include the following:

INFERTILITY

Infertility is the inability of a man or woman to produce children. If a man and woman try to have a child for twelve months without success, one or both of the partners may be infertile. The problem is thought to affect one out of ten couples in the United States.

Infertility can affect both men and women. According to the National Library of Medicine, between 30 percent and 40 percent of infertility cases are caused by "male factors." Between 40 percent and 50 percent of infertility cases are caused by "female factors." In some cases, doctors cannot determine the cause.

Many different factors can cause or contribute to infertility, including certain environmental agents. Pollutants, stress, drug use, smoking, improper nutrition, and STDs may all contribute to infertility. The three most common causes of infertility are low sperm counts, anatomic defects or disease, and blockage of the fallopian tubes.

In order to treat infertility, doctors must first determine what has caused the condition. If infertility is caused by a hormonal imbalance, for example, doctors may prescribe hormone supplements. For women with blocked fallopian tubes, doctors may recommend surgery. Couples may also choose to undergo *in vitro fertilization.* During this procedure, eggs are removed from a woman's body and fertilized with a man's sperm in a laboratory. The fertilized eggs are then implanted into a woman's uterus. In the past, babies born through in vitro fertilization were sometimes called "test-tube babies."

Pozey the Clown offers a balloon to one of 100 children born as a result of in vitro fertilization at Baystate Medical Center in Springfield, Massachusetts. The hospital began using the in vitro procedure in 1990, and the 100 children born since then and their families were invited to a celebration.

The First Test-Tube Baby

On July 25, 1978, Louise Brown, the world's first "test-tube baby," was born in Oldham, England. Louise's mother, Lesley Brown, was infertile as a result of defective fallopian tubes. To perform the fertilization procedure, doctors removed her eggs and fertilized them with sperm from her husband, John. The fertilized eggs were then implanted into Lesley's uterus. Nine months later, the couple became the proud parents of a healthy baby girl.

Although some people criticized the technique as "immoral," others welcomed the development as a medical breakthrough that allowed infertile couples the chance to become parents. Since Louise's birth, thousands of babies have been born using in vitro fertilization.

SPONTANEOUS ABORTION

A *spontaneous abortion* occurs when a fetus younger than twenty weeks of age is prematurely expelled from the uterus as a result of natural causes. This condition is often known as a miscarriage. Approximately one out of every ten known pregnancies will end this way, usually between the seventh and twelfth week of pregnancy. However, experts estimate that as many as half of all fertilized eggs will be spontaneously aborted, many before a woman even knows that she is pregnant.

Most spontaneous abortions occur after the fetus has died inside the womb, usually as a result of genetic abnormalities in the fetus itself. Other causes include infections and hormonal imbalances, as well as a woman's physical and medical conditions that may affect the fetus. Women over the age of thirty-five, those who have certain health conditions, and those who have had three or more previous miscarriages are at a higher risk for spontaneous abortion.

Symptoms of spontaneous abortion include abdominal cramps, bleeding from the vagina, and pain in the lower back. In some cases, the miscarriage can be prevented if the woman seeks immediate medical treatment. Complications of a spontaneous abortion may occur if some of the dead fetal tissue remains inside the womb. Then surgery must be performed to remove the remaining tissue. After a woman has suffered a spontaneous abortion, doctors often recommend that she wait a few months before trying to become pregnant again.

STILLBIRTH

Stillbirth is a condition in which a fetus that is more than twenty weeks of age dies before or during birth. According to the March of Dimes, about one in 200 pregnancies ends in stillbirth. The condition may be caused by any number of factors, including illness, infections, trauma (such as a motor-vehicle accident), or a genetic abnormality in the fetus. In one-third of all stillbirths, the cause is never discovered. Stillbirths are usually diagnosed during an ultrasound examination. An *ultrasound* uses high-frequency sound waves to create a picture of the inside of the body. The mother is then given medicines to expel the fetus from her body.

Sudden Infant Death Syndrome

In the United States each year, seemingly healthy babies die suddenly and mysteriously, usually as they sleep in their cribs. *Sudden infant death syndrome* (SIDS) is the third leading cause of death among children under the age of one. So far, experts have been unable to determine what causes the deadly disorder. Doctors do know that 90 percent of SIDS deaths occur in infants less than six months old. In addition, African American and Native American babies are at a higher risk of the condition. Other risk factors include premature birth, smoking or drug use by the mother, poor prenatal care, poverty, and teenage mothers.

In recent years, experts have tried many different ways to reduce the number of deaths that result from SIDS. In 1992, doctors began recommending that parents put infants to sleep on their sides or backs, not on their stomachs. Since then, the rate of SIDS in the United States has dropped by almost half. Until researchers learn more about what causes this condition, however, a complete cure will remain a mystery.

BREAST DISORDERS

Several conditions can affect the breasts. These conditions range from such harmless conditions as fibroadenoma to such life-threatening medical problems as breast cancer.

BREAST CANCER

Breast cancer occurs when cancerous tissue begins to grow in any part of the breast. It is one of the most common types of cancer in women. In fact, one out of every eight women will develop cancer of the breast at some point in her lifetime.

Although breast cancer can affect both women and men, more than 99 percent of all breast cancer patients are female. Women most at risk for breast cancer are those over the age of fifty with a history of this type of cancer in their families and a genetic predisposition to the disease. Other factors that increase a woman's chances of getting breast cancer are the use of oral contraceptives, hormone replacement therapy, obesity, alcohol abuse, exposure to certain chemicals, and radiation.

More than three-fourths of all breast cancers begin in the ducts that carry milk to the nipples. Other cases may begin in the tissue, skin, fat, and other parts of the breast. Cancerous cells can spread from the breast to other parts of the body through the blood or lymphatic (lim-FAT-ik) systems.

Symptoms of breast cancer include a lump or mass (usually painless) within the breast or the armpit, a change in the shape or size of a breast or in the appearance of a nipple, abnormal discharge from the nipple, skin changes on the breast, particularly the nipple area; and breast pain or discomfort. Symptoms of advanced breast cancer include bone pain, abnormal weight loss, and *ulcers* (open sores) of the skin.

Breast-cancer survivors and supporters celebrate at the summit of 12,460-foot (3,738-meter) Mount Fuji in Japan in August 2000. Over 200 cancer patients from the United States and Japan climbed Mount Fuji to inspire hope and determination to conquer their illnesses.

STAGING AND TREATING BREAST CANCER

When doctors talk about how severe a person's breast cancer is, they identify five different stages of the disease. Here is how breast cancer is staged:

- **STAGE 0:** breast cancer involves cancerous cells that are still in their original location in the breast and have not spread.
- **STAGE I:** breast cancer involves a cancerous tumor that is less than 2 centimeters (0.8 inches) in size and has not spread.
- **STAGE II:** breast cancer involves a larger tumor that has not spread or a tumor that has spread to one, two, or three underarm lymph nodes on the same side as the breast cancer.
- **STAGE III:** breast cancer involves a smaller-sized tumor that has spread to four or more underarm lymph nodes or a larger tumor that has spread to any underarm lymph nodes. Stage III cancer can also involve a tumor that has grown into the chest wall or skin or has spread to lymph nodes under the collarbone, above the collarbone, or near the breastbone.
- **STAGE IV:** breast cancer involves cancer that has spread to other organs beyond the breast and chest wall.

The earlier breast cancer is detected, the better the chances are of successfully treating the condition. Most cases of Stage IV breast cancer cannot be cured. For Stage I, Stage II, or Stage III breast cancer, doctors may use surgery, radiation, chemotherapy, or hormone treatment to try to cure the cancer and prevent recurrences. Most women receive a combination of treatments. Surgeries to treat breast cancer include a *lumpectomy* (lump-EK-tuh-mee), also known as a partial mastectomy, which is a procedure to remove the cancerous mass from the breast. Another procedure, the *mastectomy* (mass-TEK-tuh-mee), involves removing an entire breast. During a *radical mastectomy,* doctors will also remove the lymph nodes in the armpit that drain the affected breast. In cases where a woman is at high risk of breast cancer, both breasts may be removed at the same time.

FIBROADENOMA

A *fibroadenoma* (fye-broh-ad-NOH-muh) is a benign mass in the breast. These masses, made up of glandular tissue and fatty tissue, usually retain their size and shape. Fibroadenomas are the most common type of lumps to occur in women's breasts. They are usually found in women under the age of thirty.

Although no one knows exactly what causes these masses, some researchers believe that high-fat diets may raise a woman's risk of developing the condition. To be certain that a breast lump is indeed a fibroadenoma and not something more serious, doctors can perform a biopsy on the lump. The usual treatment for fibroadenomas is to surgically remove them from the breast.

Fast Fact

As many as 85 percent of all breast lumps are *benign,* or noncancerous.

MISCELLANEOUS CONDITIONS

Other reproductive system disorders include the following:

IMPOTENCE

Impotence, also called erectile dysfunction, is a condition in which men are unable to achieve and maintain an erection. The condition is a common one in the United States, affecting between 2 million and 30 million men each year. It is estimated that more than half of all men between the ages of forty and seventy experience problems with impotence each year.

There are two types of impotence, primary and secondary. *Primary impotence,* a rare condition, occurs when a man has never been able to achieve and maintain an erection. *Secondary impotence* occurs when a man loses the ability to achieve and maintain an erection. This loss of ability usually occurs gradually. Most cases of impotence are secondary.

Nine out of ten cases of impotence are caused by physical conditions, including medications, high blood pressure, diabetes, and liver or heart disease. Environmental agents that can contribute to the condition include smoking, alcohol use, and illegal drug use.

Although the risk of developing impotence increases with age, aging by itself does not cause the condition.

In order to treat impotence, doctors must first determine what has caused it. If a particular medication is causing the problem, for example, doctors may recommend an alternative medication that will not have the same side effects. If impotence is caused by smoking, alcohol use, or other lifestyle choices, doctors will recommend that the patient seek other, healthier activities. There are also medications and surgical procedures that can reverse the problem.

MENSTRUAL DISORDERS

Women can experience a number of problems with the menstrual cycle. *Primary amenorrhea* (ay-men-uh-REE-uh) is a condition in which a female does not begin menstruating by the age of sixteen. This condition may be caused by a number of factors, including an imbalance of hormones, tumors on the pituitary gland, hypoglycemia, anorexia (an-uh-REKS-ee-uh), obesity, and other medical conditions. Doctors treat the condition depending upon what is causing it. If amenorrhea is caused by a hormone imbalance, for example, doctors will prescribe hormone supplements. Pituitary tumors are usually removed by surgery.

Secondary amenorrhea occurs when a person who has previously menstruated suddenly stops doing so. The condition may occur as a result of stress and anxiety, too much exercise, obesity, extreme thinness, and other medical conditions. About 4 percent of American women are affected by secondary amenorrhea. Treatment of the condition varies, depending upon the cause.

Dysmenorrhea (diss-men-uh-REE-uh) is a condition in which menstrual cycles are painful or uncomfortable. While some abdominal cramping is normal just before a woman's period, severe pain may indicate a more serious medical condition, such as pelvic inflammatory disease (PID), a

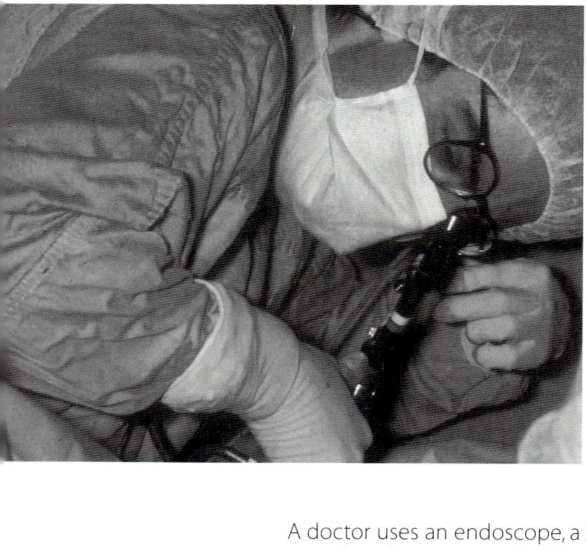

A doctor uses an endoscope, a slender tube with a light source through which a physician can examine internal organs, to diagnose menstrual disorders.

condition usually caused by STDs, which scar the fallopian tubes. To ease the symptoms of dysmenorrhea, doctors may recommend over-the-counter painkillers, heating pads placed over the abdomen, warm beverages, or walking.

Menorrhagia (men-uh-RAY-jee-uh), also known as hypermenorrhea (hye-per-men-uh-REE-uh), is a condition characterized by extremely heavy periods. During a normal period, a woman loses between 1 tablespoon and 1 cup (15 to 240 milliliters) of blood. Periods that are heavier may be the sign of other medical conditions, including some types of cancer, uterine tumors, abnormal endocrine functioning, or miscarriage. Women with menorrhagia should consult a doctor. Doctors may recommend bed rest, painkillers, or hormone therapy to treat the condition. In some cases, a surgical procedure may be performed in which doctors scrape the uterine lining. The lining is then examined under a microscope for abnormalities.

PREMENSTRUAL SYNDROME

One of the most common conditions associated with menstruation is *premenstrual syndrome* (PMS). Experts believe that between 70 and 90 percent of all menstruating women suffer from some of the symptoms of PMS. Of these women, 30 percent to 40 percent are severely affected by the condition. Although doctors aren't sure what exactly causes PMS, some believe that hormonal changes in the body during the menstrual cycle may be the culprit. PMS most often affects women between five and ten days before their periods begin.

PMS can cause both physical and emotional symptoms. Physical symptoms of PMS include cramps, bloating, acne, aching breasts and back, headaches, and diarrhea. The emotional effects of the condition include irritability, depression, food cravings, and difficulty concentrating. Symptoms usually disappear once menstruation begins.

Researchers say that the best cures for PMS are safe and easy—exercising, eating a healthy diet, getting plenty of rest, and avoiding alcohol, tobacco, and caffeine to ease symptoms. Some health experts also believe that these healthy activities can prevent women from ever developing PMS. For severe cases, doctors may recommend over-the-counter painkillers, vitamin and hormone supplements, or prescription medications.

CERVICAL CANCER

Cervical cancer is any cancer of the cervix. It is the third most common type of cancer among women throughout the world. Cervical cancer develops slowly, starting off as the growth of abnormal, precancerous cells in the cervix. This abnormal growth is known as *dysplasia* (diss-PLAY-zhee-uh). Dysplasia can be detected through a routine Pap smear.

Doctors are not sure what causes cervical cancer. However, many factors increase the chances of a woman getting the disease. These risk factors include having sexual intercourse before the age of seventeen, having multiple sexual partners, smoking, using birth control pills for five years or more, and having a diet that is low in vitamin A. Women who have a history of sexually transmitted diseases are also at higher risk.

> **Fast Fact**
>
> Women in the United States are less likely to be diagnosed with cervical cancer than are women in other countries. This is because American women are more likely to receive annual Pap smears, which detect precancerous growths early.

In the early stages of cervical cancer, there are few symptoms. Later, symptoms may include abnormal or foul-smelling discharge and abnormal vaginal bleeding. As the cancer advances, symptoms include pelvic, back, or leg pain; loss of appetite; weight loss; fatigue; the leaking of urine or feces through the vagina; and bone fractures.

Doctors treat cervical cancer based on its stage. In early stages, precancerous cells can usually be successfully removed from the cervix. In later stages, a woman must have a hysterectomy (hiss-teh-REK-tuh-mee), a procedure in which a woman's entire uterus is removed. Radiation and chemotherapy may also be used to treat the cancer.

PROSTATE CANCER

Prostate cancer is a condition in which a cancerous mass grows within the prostate gland. Prostate cancer is the third most deadly type of cancer among men of all ages in the United States. It is the most deadly form of cancer among men over the age of seventy-five.

Researchers aren't sure what causes prostate cancer. They do know that this type of cancer is more likely to affect men over the age of fifty. They also know that men who have a family history of the disease, as well as African American men, are more at risk for prostate cancer. In addition, some studies indicate a link between prostate cancer and high-fat diets or abnormal testosterone levels in the body. Prostate cancer also more commonly affects male workers who have been exposed to cadmium, an element used to make solder, dental fillings, batteries, and other items.

Symptoms of prostate cancer include difficulty urinating, pain while urinating, lower back pain, and pain when having a bowel movement. Secondary symptoms include excess urination at night, abdominal pain, aching or painful bones, and blood in the urine. Treatment of the cancer can include surgical removal of the prostate gland, chemotherapy, radiation therapy, and hormone treatment to regulate testosterone levels in the body.

TESTICULAR CANCER

Testicular cancer is a condition that occurs when abnormal cells in the testicles rapidly multiply. Each year, as many as 8,000 American men are diagnosed with testicular cancer. It is the most common type of cancer in men between the ages of fifteen and forty. Doctors are not sure what causes this type of cancer. They do know, however, that men with certain medical conditions, including abnormal testes and other testicular disorders, face a higher risk than others. White males are more often at risk than are African American or Asian American men.

Symptoms of testicular cancer include an enlargement, lump, or swelling in a testicle, aching in the back or lower abdomen, excessive development of breast tissue, and pain in the testicles. However, some men may experience no symptoms at all.

Treatment of testicular cancer depends upon which of three stages the cancer is in. Stages are the way doctors describe the severity of a cancer case. Stages of testicular cancer range from Stage I, the least serious, in which the cancer is limited to the testicles only, up to Stage III, in which the cancer has spread to other organs. Treatment usually consists of surgery, radiation, chemotherapy, or a combination of these treatments. In early stages, the usual treatment is to remove the testicle that contains the cancerous mass.

Lance Armstrong, Cancer Survivor

As the world's greatest cyclist, Lance Armstrong (1971–) has faced many challenges and uphill battles, but the toughest fight of Armstrong's life was against testicular cancer. In 1996, twenty-five-year-old Armstrong was diagnosed with Stage III testicular cancer. Because he had ignored the warning signs, the cancer had time to spread to his abdomen, lungs, and brain. Doctors told the young man that his life was at stake.

Armstrong underwent surgery to remove the cancerous testicle, as well as cancerous tumors in his brain. He was also treated with chemotherapy. A year later, Armstrong was nearly cancer-free. It wasn't long before he was back on his bike, stronger and more determined than ever before.

In 1999, Armstrong inspired cancer survivors everywhere when he won the prestigious Tour de France, a bike race in France. He followed up on his victory by winning the Tour in 2000, 2001, and 2002. When Armstrong isn't racing, he oversees the Lance Armstrong Foundation, an association that provides support, education, and advocacy for cancer patients and survivors around the world.

Testicular cancer has one of the highest cure rates of any type of cancer. The earlier the cancer is detected, the better the chances of successfully treating the condition. For this reason, it is important for young men to learn how to perform monthly exams to identify potential problems early. When detected early enough, some types of testicular cancer have a 95 percent cure rate.

Self-Exams for Early Cancer Detection

Some kinds of cancer may be detected early by the person with the tumor. Breast cancer is one such type. Doctors recommend that, each month, women over the age of twenty examine their own breasts to look for any new lumps that might be developing. It is important to perform the test each month to learn what each breast normally feels like, so abnormalities will be easier to detect. If a woman detects a breast lump, she should make an appointment with her doctor to get the lump checked out.

Men can perform self-exams by feeling their testicles for lumps or swellings that may be caused by testicular cancer. Doctors recommend that men perform these exams right after a bath or shower. Any new lumps or abnormalities should be reported to a physician immediately.

DIAGNOSING REPRODUCTIVE DISORDERS

Doctors use a number of tests to help them diagnose reproductive system disorders. Imaging tests such as MRIs, CT scans, and ultrasounds can be used to look inside the body for evidence of such conditions as tumors, cysts, and cancerous masses. One special type of imaging test used to detect breast cancer and other breast problems is the *mammogram.* During a mammogram, a woman's breast is placed on a clear plate. Pressure is applied to make sure that all the tissue is even, and an X-ray is then taken.

Biopsies may be performed for further examination. Samples may be taken with long needles or during surgery. The tissues or fluids extracted during the biopsy are examined under a microscope for evidence of abnormalities.

Blood tests for increased amounts of prostate-specific antigen (AN-tih-jen), or PSA, help doctors detect prostate cancer. PSA is a protein that is made in the prostate. High levels of PSA in the blood can indicate cancer or prostate infection.

A *Pap smear* is an important part of female reproductive health. During the test, doctors scrape tissue from the cervix. The tissue is later examined under a microscope for traces of abnormal cells. The Pap smear is a valuable tool in the early detection of cervical cancer and inflammation.

6

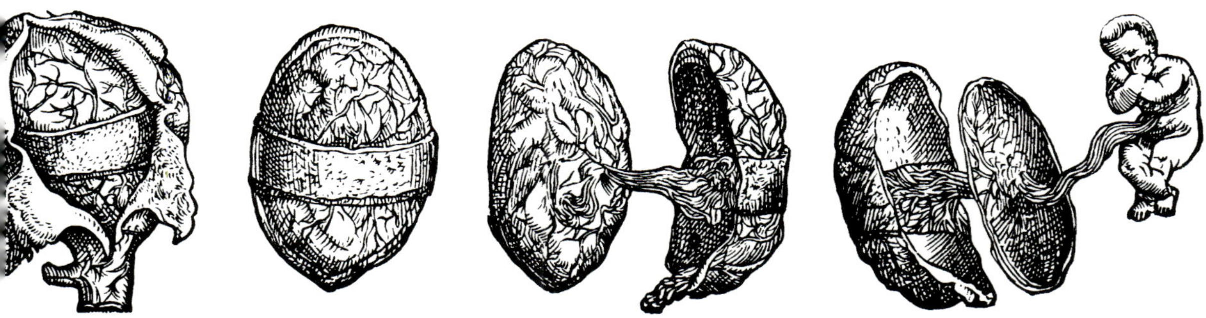

HOW THE ENVIRONMENT AFFECTS THE REPRODUCTIVE SYSTEM

Like other body systems, the reproductive system can be seriously affected by environmental agents. These agents include chemicals, radiation, bacteria, viruses, other harmful organisms, and even lifestyle choices.

CHEMICALS AND RADIATION

Every day, people come in contact with chemicals that are present in the world around them. We breathe in chemicals that pollute the air. We ingest chemicals in contaminated water and food. Some of these chemicals can cause serious damage to the reproductive system. Chemicals that have been shown to affect reproductive health include dioxin, lead, and mercury.

Unborn fetuses are especially at risk from exposure to harmful chemicals. A developing fetus is connected to its mother by the umbilical cord. Nutrients travel from mother to fetus through the cord. When a pregnant woman ingests chemicals from the environment, these harmful toxins also pass through the umbilical cord to the baby. Because the fetus is still developing, exposure to environmental toxins can result in miscarriage or serious birth defects.

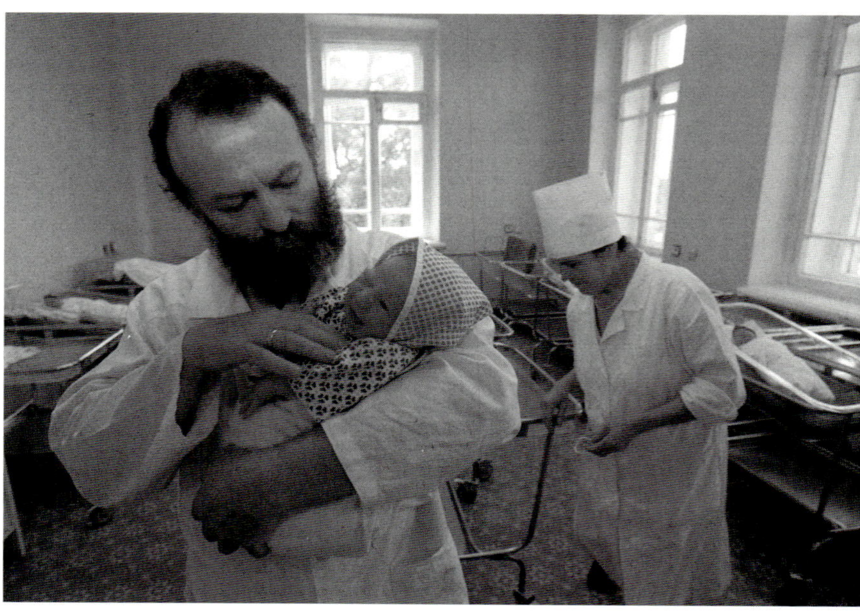

A doctor in Dzerzhinsk, Russia, holds a newborn baby. Babies born in Dzerzhinsk have triple the number of birth defects of babies born in other Russian cities, most likely because of chemicals released into the water system by factories and chemical plants located in and around the city.

One of the most serious examples of the effect of chemicals on unborn babies occurred in Canada and Europe in the early 1960s. During that time, many doctors treated pregnant women who experienced morning sickness with a drug called *thalidomide* (thuh-LID-uh-myde). The drug caused serious, disabling birth defects in developing fetuses. Women who took the drug when their infants were forming arms and legs, for example, gave birth to children whose limbs were missing or severely malformed. Other children were born with defective hearts, genitals, kidneys, and other organs. Today, thalidomide has been banned for use by pregnant women. However, the drug is still used to treat leprosy and some other medical conditions.

In recent years, a number of studies have been performed to learn more about the effects of chemicals on reproductive health. In 2002, for example, a study showed that men in farming areas of Missouri had lower sperm counts than men in city areas throughout the nation. Researchers believe that the use of chemical fertilizers or pesticides may have caused this difference.

Radiation, like chemical agents, can affect men's reproductive health. Radiation is believed, for example, to cause decreased sperm counts and infertility. Chemicals and radiation can affect the reproductive health of females, too. Women exposed to radiation, lead, and certain chemicals are at higher risk of infertility, miscarriages, and delivering babies with birth defects and other problems. They may also experience premature menopause. For this reason, pregnant women should not receive X-rays and should limit their exposure to chemicals and other potential toxins.

BACTERIA, VIRUSES, AND OTHER HARMFUL ORGANISMS

Viruses, bacteria, and fungi can cause a number of reproductive system diseases. Left untreated, these infections can lead to serious problems. Pregnant women and their fetuses are especially at risk. Such harmful organisms as the viruses that cause hepatitis B, AIDS, rubella, and chicken pox can lead to miscarriages or cause children to be born with birth defects.

BACTERIAL VAGINOSIS

Bacterial vaginosis (vaj-ih-NOH-siss) is an irritation of the vagina that is caused by bacteria. According to the Centers for Disease Control (CDC), bacterial vaginosis is the most common vaginal infection among U.S. women of childbearing age. The most common cause of this condition is the bacterium *Gardnerella vaginalis* (gard-ner-ELL-uh vaj-ih-NAL-iss). Pregnant women seem to be at especially high risk of developing the condition. According to the CDC, as many as 16 percent of all pregnant women have the condition at some point during their pregnancy.

Symptoms of the disorder include a thin, gray vaginal discharge and vaginal itching or burning. Some women, however, will exhibit no symptoms at all. Doctors treat the condition with medicines that kill the bacteria. If left untreated, bacterial vaginosis may lead to increased risk of PID and increased susceptibility to STDs.

BREAST INFECTION

Breast infection, also known as mastitis (mass-TYE-tiss), is a condition in which bacteria enter the breast, causing infection in the breast tissue. Breast infections usually occur in women who are breastfeeding a child. The bacteria causing the infection usually enter through cracks or breaks in the skin of the nipple.

Symptoms of breast infection include pain, swelling, heat, and tenderness in the affected breast, discharge (often pus) from the breast, fever, itching of the infected area, and swollen lymph nodes of the armpit nearest the infected breast. Doctors usually prescribe antibiotics to treat the infection. To ease pain and swelling, they may recommend that women apply moist heat to the affected breast several times a day.

CANDIDIASIS

Candidiasis (kan-dih-DYE-uh-siss), also known as a yeast infection, is a condition caused by an excess of *Candida albicans* (KAN-did-uh AL-bih-kanz), a yeastlike fungus. *Candida albicans* is naturally present in the vagina, mouth, and digestive tract, as well as on the skin. However, problems arise when the fungus grows too rapidly. An excess of the fungus can cause an irritation of the vulva or vagina.

Candidiasis may occur after a woman has taken antibiotics to cure a bacterial infection. Antibiotics can upset the balance of fungus-controlling bacteria inside the vagina. As a result, the fungus begins to grow out of control. Higher than normal levels of estrogen in the body can also trigger yeast infections. People with diabetes and immune system disorders such as HIV infection are more at risk for yeast infections than are other populations. In addition, wearing wet or damp clothing (such as undergarments or bathing suits) provides the fungus with an excellent environment in which to flourish.

Symptoms of candidiasis include a clear or chunky white vaginal discharge, as well as pain, itching, and burning of the vaginal area. Some patients may also experience pain when urinating. Candidiasis can usually be treated with over-the-counter medicines that are inserted into the vagina to kill fungi.

LIFESTYLE CHOICES

Everyone has choices to make in life, but some of the choices that people make can actually jeopardize their health. Engaging in unsafe sexual intercourse, for example, can lead to the development of sexually transmitted diseases. Smoking and the abuse of alcohol and illegal drugs can also cause reproductive health problems.

HIV, AIDS, and Unsafe Sex

The *human immunodeficiency* (im-myoon-oh-dih-FISH-en-see) *virus* (HIV) is a contagious virus that attacks the body's immune system. HIV can lead to AIDS, a serious, life-threatening condition. HIV is spread through contact with infected blood and other body fluids. One of the chief ways that infected fluids are transmitted from one person to another is through unprotected sexual intercourse. In fact, 75 percent of all new HIV infections in the United States are the result of sexual intercourse. Each year, about 40,000 Americans are infected with HIV.

SEXUALLY TRANSMITTED DISEASES

Half of all reported infectious diseases are *sexually transmitted diseases* (STDs). These diseases are passed from one person to another through unprotected sexual intercourse—sexual intercourse without the use of a condom—whether oral, vaginal, or anal. According to the United States Department of Health and Human Services, about 12 million cases of STDs are diagnosed in this country each year. One-quarter of these cases are diagnosed in teenagers. While some STDs are curable, others cannot be cured, leading to problems that last for the rest of a person's life. Common STDs include the following:

CHLAMYDIA

Chlamydia (kluh-MID-ee-uh) is an STD caused by the bacteria *Chlamydia trachomatis* (truh-KOH-muh-tiss). In the United States, chlamydia is the most common of all bacterial STDs. According to the CDC, more than 650,000 cases of chlamydia were reported in the United States in 1999. Three-fourths of these cases were diagnosed in people who were younger than twenty-five.

Chlamydia infection can affect the urethra, uterus, rectum, and throat. Symptoms of this STD include genital discharge, abdominal and genital pain, a burning sensation during urination, and pain during sexual intercourse. However, many patients experience no symptoms at all and may not know that they have contracted the disease. In fact, 70 percent of women and 25 percent of men infected with chlamydia will be *asymptomatic* (ay-simp-toh-MAT-ik), or without symptoms.

Chlamydia can usually be treated successfully with antibiotics. If this STD is left untreated, it can affect the fallopian tubes, causing PID. PID, in turn, can lead to an increased risk of infertility.

GENITAL HERPES

Genital herpes (HER-peez) is an STD that is caused by a virus, herpes simplex virus type 2 (HSV-2). (Another strain of this virus, HSV-1, causes cold sores around the mouth.) According to the National Library of Medicine, about 86 million people worldwide have genital herpes. In the United States, the CDC estimates that about one of every five people age twelve and older is infected with the virus.

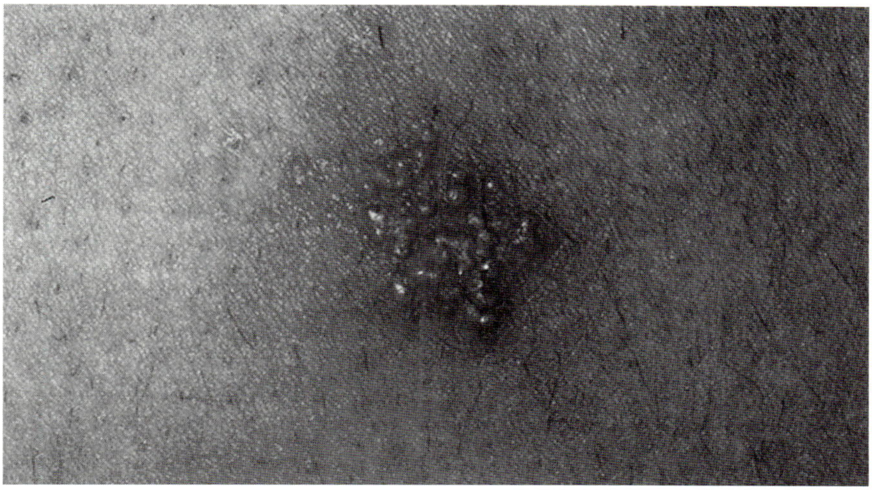

Genital herpes is characterized by blisters and inflammation, as shown in this photograph.

Symptoms of the condition include painful, fluid-filled blisters that recur periodically in the genital area. These blisters eventually burst, leaving behind ulcers that heal gradually over one to two weeks. Other secondary symptoms include vaginal discharge in women and painful urination, fever, muscle pain, and loss of appetite in both men and women. A pregnant woman who suffers from herpes may pass the virus to her fetus, causing health problems and even death. People with immune-system disorders may suffer serious complications from genital herpes, including infections of the eyes, nose, throat, esophagus (ih-SAHF-ih-guss), liver, brain, and lungs.

Because there is currently no cure for genital herpes, the condition may recur throughout a patient's life. There are medicines available, however, to ease the pain associated with the disease and to speed up the healing process.

GENITAL WARTS

Genital warts are caused by a virus called the *human papillomavirus* (pap-ih-LOH-muh-vye-russ), or HPV. Symptoms of the condition include itching, burning, pain, and fleshy, wartlike growths in the genital area. These growths can affect the vagina, uterus, cervix, urethra, penis, and rectum. Most people, however, exhibit no symptoms at all. Scientists have identified more than 100 different types of genital warts. According to the CDC, HPV can be transmitted from one sexual partner to another even when a condom is used.

Genital warts cannot be cured. However, the condition should be treated in order to avoid complications. Without medial attention, genital warts can increase in size. They can spread into the cervix, increasing the risk of cancer of the cervix and vulva. Treatments for the condition include topical medications that can be applied to the warts or surgery to remove them.

GONORRHEA

Gonorrhea (gahn-er-EE-uh) is an STD caused by the bacteria *Neisseria gonorrhoeae* (nye-SEER-ee-uh gahn-er-EE-ay). Gonorrhea is one of the most common of all STDs. Each year, nearly 400,000 new cases of the disease are reported to the CDC. Some health experts believe that many other cases go unreported. For this reason, some experts estimate that as many as 2 million Americans are affected by gonorrhea each year.

Symptoms of gonorrhea include clear, white, or yellow genital discharge and a painful burning sensation during urination. The condition may also cause throat, rectal, and eye infections. However, like chlamydia, gonorrhea may cause no symptoms at all, so some patients don't even know that they have contracted the disease. When gonorrhea is detected early enough, doctors can usually treat it successfully with antibiotics. Left untreated, however, gonorrhea can cause serious health complications, including narrowing and scarring of the urethra, which may eventually lead to kidney failure.

PELVIC INFLAMMATORY DISEASE

Pelvic inflammatory disease (PID) is a medical condition that involves an infection of the fallopian tubes, uterus, or ovaries. The condition is usually caused by an STD, although it may also be caused when bacteria enter the body during childbirth, abortion, or other procedures that relate to women's reproductive health. The condition is common in the United States, affecting one million females each year. Those at higher risk are women who were sexually active as adolescents, those who have had multiple sexual partners, and those who have had an STD or a prior case of PID.

Symptoms of PID include an abnormal vaginal discharge, abdominal pain, and fever. Secondary symptoms include irregular menstrual periods, back pain, fatigue, nausea, and pain when urinating. Doctors usually treat PID with antibiotics to kill the bacteria that are causing the condition. In serious cases, the patient may need to be hospitalized so that doctors can treat widespread infections with intravenous antibiotics. Complications of the condition include an increased risk of infertility.

SYPHILIS

Syphilis (SIF-ih-liss) is an STD caused by the bacteria *Treponema pallidum* (treh-poh-NEE-muh PAL-ih-dum). Syphilis is most often transmitted through sexual intercourse. The disease has three stages: primary, secondary, and tertiary. Primary syphilis is the earliest stage of the disease. During this stage, the patient develops painless blisters between three to six weeks after infection. These blisters are known as *chancres* (SHANK-erz). Primary syphilis can almost always be successfully treated and cured with antibiotics. If left untreated, chancres usually disappear. The syphilis bacteria, however, remain active within the body, continuing to cause damage.

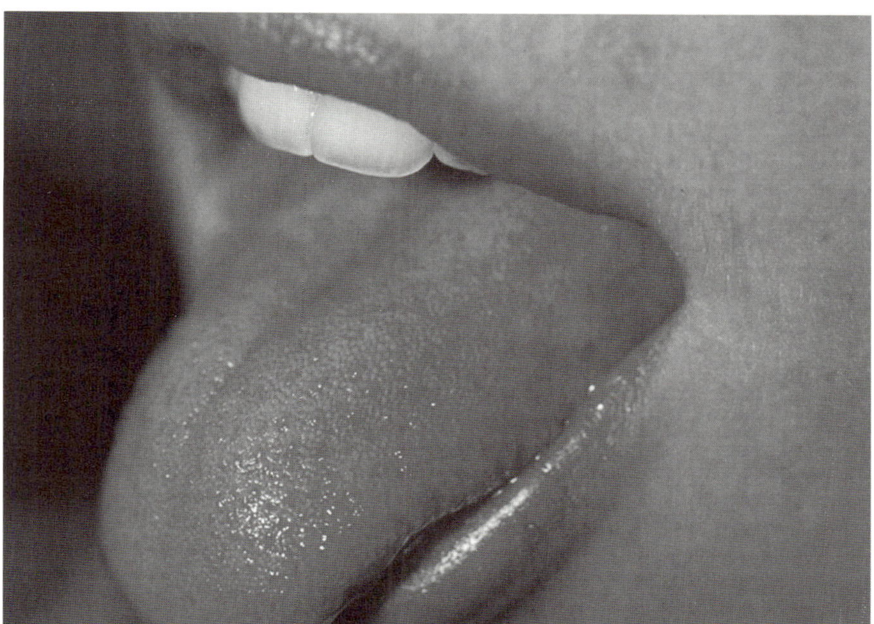

Secondary syphilis can cause sores on the mouth and tongue as well as a rash on the palms of the hands and soles of the feet.

Two to eight weeks after the chancres first appear, the patient may develop a rash in the genital area. This rash, which may also affect the palms and soles of the feet, is the first sign of secondary syphilis. Other symptoms of this stage include sores on the mouth and genitals, fever, and swollen lymph nodes. Secondary syphilis is the most contagious stage of the disease. Like primary syphilis, secondary syphilis is treated with antibiotics, usually penicillin (pen-uh-SILL-in). Without treatment, secondary syphilis can last anywhere from a few weeks to a full year.

If a person with secondary syphilis does not seek treatment, he or she may develop the third stage of the disease, known as tertiary, or late, syphilis. Tertiary syphilis, which may take between three and fifteen years to develop, is characterized by health complications resulting from the spread of bacteria to such organs as the brain, bones, kidneys, and heart. Here, the bacteria can cause serious damage and even lead to death. This stage of the disease is rare today, thanks to early detection and antibiotic treatment.

Reportable Diseases

Doctors and health professionals in all 50 states are required by law to report certain diseases to their state health authorities. When these diseases are reported, doctors are able to make sure that a patient receives follow-up care. Reporting such diseases also allows health officials to notify the patient's sexual partners that they may have been exposed to an infectious disease. According to the National Library of Medicine, as many as 90 percent of an infected male's sexual partners will be infected with his disease. Reportable STDs include chlamydia, gonorrhea, HIV infection, and syphilis. However, not all reportable disease are sexual in nature. Other reportable diseases include cancer, Legionnaire's disease, measles, and whooping cough.

Pregnant women with syphilis place their babies at risk of infection with the disease, as well as other medical complications. When infants are infected by their mothers, the condition is known as *congenital* (kun-JEN-ih-tul) *syphilis.* Congenital syphilis is a life-threatening condition. About half of all fetuses who are infected with the disease die before or shortly after birth. Babies who do survive may suffer such symptoms as failure to grow, irritability, bone lesions, facial deformities, and skin rashes. Later, they may become blind or deaf.

Because of the harm that congenital syphilis can cause, pregnant women are routinely tested for syphilis early in their pregnancies. If they are found to have the disease, doctors treat them with penicillin. This lowers the risk that babies will be born with congenital syphilis.

ALCOHOL AND ILLEGAL DRUGS

The use of alcohol and illegal drugs can have serious effects on the human reproductive system. In addition, alcohol abuse and illegal-drug use can lead people to make risky decisions, such as engaging in unsafe sex.

FETAL-ALCOHOL SYNDROME

Fetal-alcohol syndrome is a medical condition in children caused when their mothers drank alcohol while they were pregnant. Alcohol can cause birth defects, including mental retardation, heart damage, musculoskeletal (mus-kyoo-loh-SKEL-uh-tul) problems, and other disorders. The more alcohol an expectant mother drinks, the higher the risk of these problems for her child.

Researchers are not sure how much alcohol consumption will cause birth defects. They do know that alcohol remains longer in a developing fetus's system than the mother's system. Health professionals recommend that pregnant women not consume any alcohol until after they give birth (or after they have stopped breastfeeding).

Symptoms of fetal-alcohol syndrome include failure to grow, delayed development and signs of mental retardation, facial abnormalities, and excessive crying. Further testing may reveal serious heart problems.

Women who consume alcohol during pregnancy also raise their chances of miscarriage or stillbirth. Babies born to mothers who drank during their pregnancy are at higher risk of low birth weight and death during early infancy.

COCAINE AND UNBORN BABIES

According to the Center for Evaluation of Risks to Human Reproduction (CERHR), one of the most dangerous drugs for unborn fetuses is cocaine. Cocaine is an illegal stimulant. When used by pregnant women, it can cause miscarriages, premature labor, and stroke or death in the unborn baby. It doubles the risk of a woman delivering her baby prematurely. Cocaine can also cause the placenta to rip away from the uterus, leading to severe internal bleeding. (The placenta is an organ by which the fetus is attached to its mother's uterine wall.) This bleeding puts both mother and child at risk of death.

Each year, thousands of American babies are born after having been exposed to cocaine. Many children born with cocaine in their system are smaller than normal. These small, low-birth-weight babies are more likely to die in infancy. Babies born to mothers who used cocaine are also more likely to have some types of birth defects, including brain damage, and feeding and sleep problems.

Problems for cocaine-exposed infants continue as those babies grow older. A 2002 study noted that two-year-olds who had been exposed to cocaine as fetuses were twice as likely as other toddlers to suffer delays in mental skills.

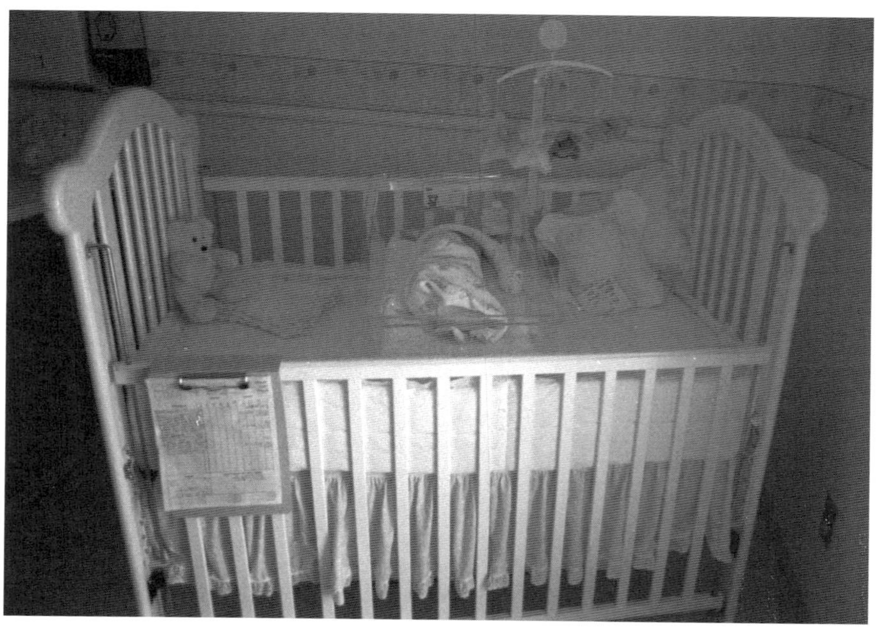

A newborn, drug-addicted baby rests in her crib at the Pediatric Interim Care Center in Kent, Washington. Babies who are addicted to the drugs their mothers took during their pregnancies are kept in dark rooms, protected from bright lights and noise that would overstimulate them.

7

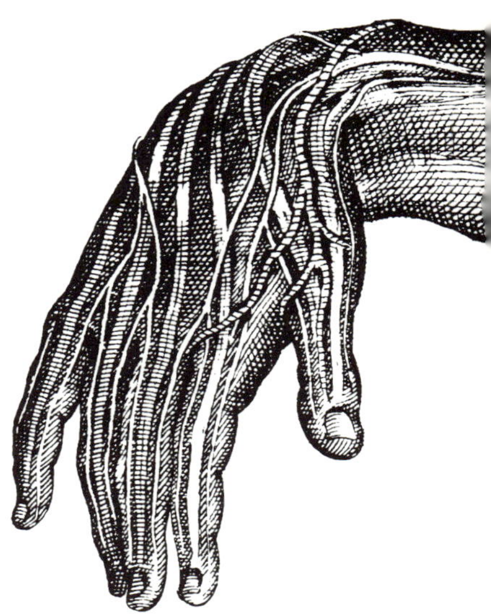

THE NERVOUS SYSTEM

The nervous system is the master-control system of the human body. Made up of the brain, spinal cord, and a network of nerve cells called neurons, the nervous system carries and decodes messages from around the body. Messages from the organs and tissues are sent, via nerve cells, to the brain, where they are translated and acted upon. Every action that people take, every function that people perform, is controlled by the nervous system. This includes ***conscious actions*** that people choose to take, such as talking, eating, walking, and running, as well as ***unconscious actions,*** such as breathing, kidney function, heartbeat, and digestion, that are automatic.

The nervous system is sometimes described as two separate sections, the central nervous system and the peripheral nervous system. The ***central nervous system*** is made up of the brain and spinal cord. The ***peripheral*** (pur-IFF-er-ul) ***nervous system*** is defined as the nerves outside the central nervous system. These nerves relay messages from the rest of the body to the brain.

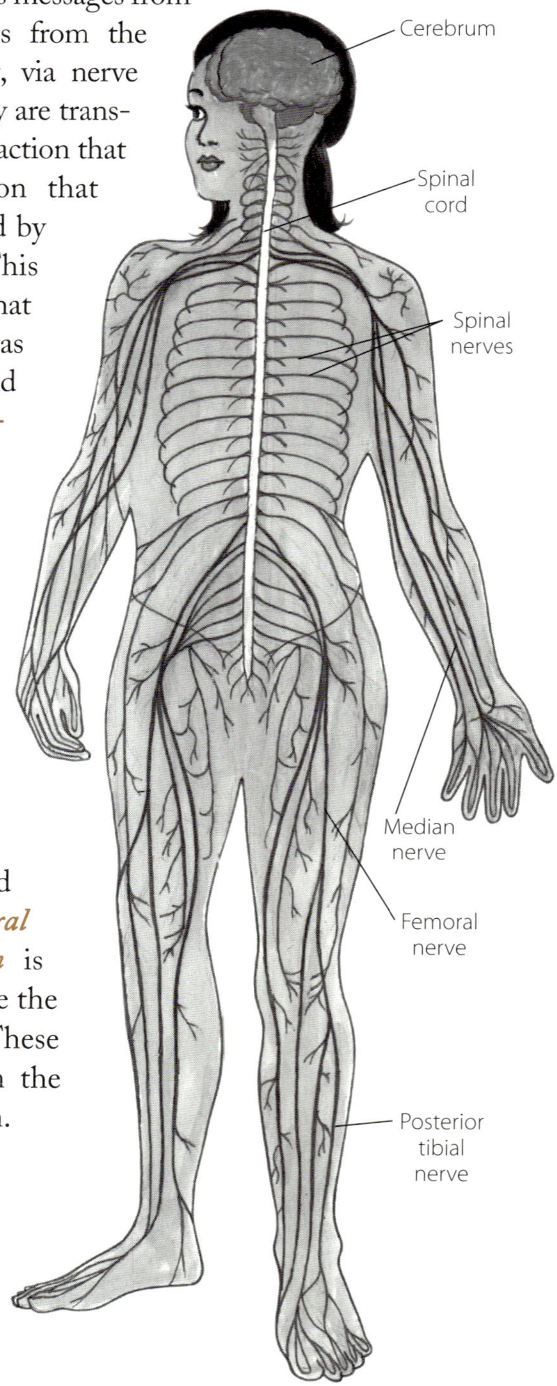

The illustration shows the parts of the nervous system distributed throughout the body.

NEURONS

Neurons, or nerve cells, are the nervous system's messengers. These special cells carry messages to the brain from every part of the body. Each nerve cell is made up of a cell body, many dendrites, and one axon. The dendrites receive incoming messages, while the axon transmits messages. There are many different types of neurons in the body, including *sensory neurons,* which carry messages between the sense organs and the brain, and *motor neurons,* which carry messages between muscles and the brain.

THE SPINAL CORD

Messages between the nerves and the brain are transported along the *spinal cord,* the body's superhighway. The spinal cord is a tube of nerve tissue about 18 inches (45 centimeters) long that runs from the base of the brain down the center of the back. The spinal cord is surrounded by cerebrospinal (suh-ree-broh-SPYE-nul) fluid, which protects the spinal cord and keeps it moist. This fluid also runs through the center of the spinal cord. Within the spinal cord are thirty-one pairs of nerves. Each pair receives messages from and transmits messages to a certain area of the body.

THE BRAIN

The *brain* is the body's main control center. Here, messages are received and translated. The brain responds to these messages, sending out commands to the rest of the body. The brain is housed in the cranium, the part of the skull that protects this important organ. Between the brain and the skull are three protective membranes, called the meninges (men-IN-jeez). The innermost layer of the meninges is made up of many blood vessels that extend deep into the brain. These blood

Fast Fact

The nervous system processes two different types of messages: afferent and efferent. *Afferent messages* are the messages sent to the brain from the rest of the body. *Efferent messages* are those sent from the brain to the rest of the body.

vessels provide oxygen to the brain. Cerebrospinal fluid bathes the brain, keeping it moist and providing it with extra cushioning.

The brain itself is made up of two types of tissue, gray matter and white matter. *Gray matter* consists entirely of nerve-cell bodies, while *white matter* is made up of axons and dendrites. The brain has several different sections, including the cerebrum, cerebellum, brain stem, and limbic system.

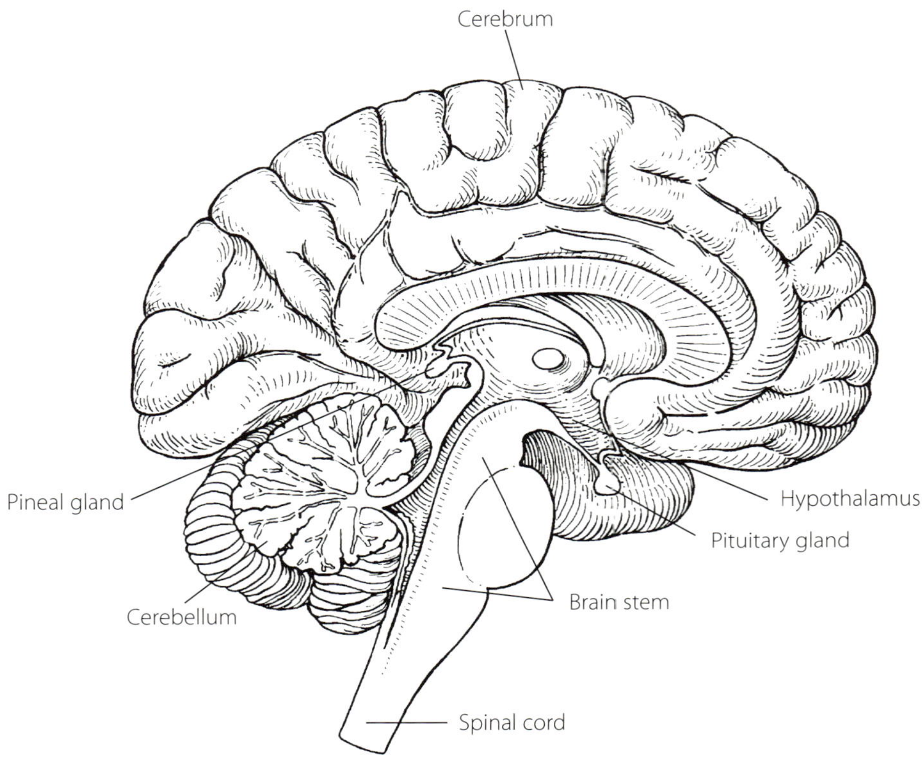

- The *thalamus* is the part of the brain through which all sensory messages pass.
- The *hypothalamus* processes information from the body concerning hunger, thirst, and sleep. It also regulates hormone production in the pituitary gland.
- The *pituitary gland,* sometimes called the "master gland," manufactures activating hormones that signal the endocrine glands to produce other hormones.

THE CEREBRUM

The *cerebrum* (suh-REE-brum) is the largest part of the brain. It makes up about four-fifths of the organ's entire weight. The cerebrum is divided into two sections, the right and left cerebral (suh-REE-brul) hemispheres. These two hemispheres are covered by a lining of cells called the *cerebral cortex.* It is the cerebral cortex that gives the brain its gray, grooved appearance. Each side of the cerebrum is made up of four different sections, known as lobes. These lobes are the frontal, parietal, temporal, and occipital. Each lobe is responsible for different functions in the body.

- The *frontal lobe* is responsible for complex thoughts and decision making. Messages to move the voluntary muscles are also sent from this part of the brain.
- The *parietal* (puh-RYE-eh-tul) *lobe* processes some types of sensory information.
- The *temporal lobe* processes written and spoken-language information.
- The *occipital* (ahk-SIH-pih-tul) *lobe* processes visual messages from the eyes.

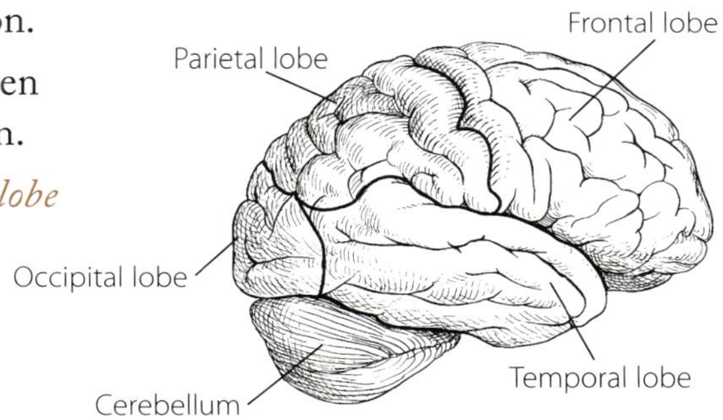

Fascinating Brain Facts

- The normal adult brain weighs about 3 pounds (1.35 kilograms).
- The brain is made up of about 100 million nerve cells that receive and transmit messages to and from the rest of the body.
- The left side of the cerebrum controls the right side of the body. The right side of the cerebrum controls the left side of the body.
- Some experts believe that the right side of the cerebrum is in charge of abstract thinking, such as processing colors, shapes, and music. These experts believe that the left side controls more concrete thoughts, such as processing mathematics, logic, and speech. That's why you'll sometimes hear creative people referred to as "right-brained."

THE CEREBELLUM

Located below the cerebrum, the *cerebellum* (sayr-uh-BELL-um) controls body movement and balance. Like the larger cerebrum, the cerebellum is divided into two separate sides.

THE BRAIN STEM

The *brain stem* is located in front of the cerebellum. Its main job is to connect the rest of the brain to the spinal cord. It also controls involuntary responses, such as breathing, circulation, and digestion. The brain stem includes the midbrain, medulla oblongata, and pons.

- The *midbrain* controls automatic functions of the eye (dilation of the pupil, focusing of the lens) and other sensory functions.
- The *medulla oblongata* (meh-DUL-uh ah-blong-GAH-tuh) is the long stalk of the brain stem that connects the brain with the spinal cord.
- The *pons* is a bundle of nerve fibers that connects the medulla oblongata to the cerebrum. It also connects the two sides of the cerebellum to each other.

THE LIMBIC SYSTEM

The *limbic system* is the parts of the brain that process emotions, memories, smell, and sexual desire. Important parts of the limbic system are the thalamus, hypothalamus, and pituitary gland.

8

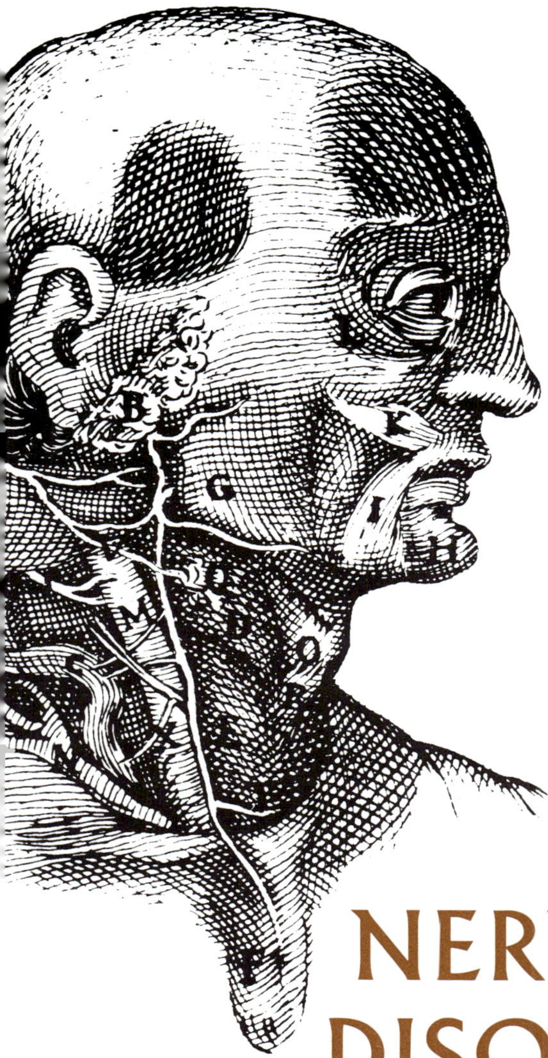

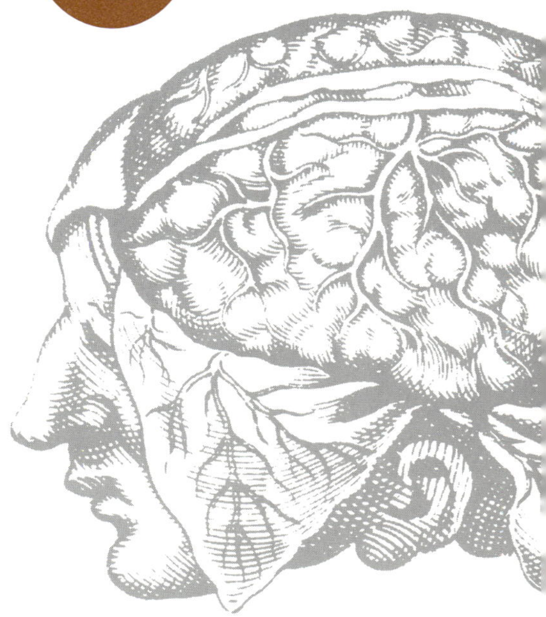

NERVOUS DISORDERS

When trauma or disease damages the brain and other parts of the nervous system, the entire body may be affected. Because parts of the brain control different behaviors, trauma to one area of the brain may affect those behaviors. Nervous-system disorders can result in tremors, paralysis, headaches, and the impairment of other bodily functions.

AGING CHANGES

Like other body systems, the nervous system can be affected by the aging process. As people age the weight of the nerve cells, brain, and spinal cord decreases. Other changes in the nerve cells cause messages to be transmitted to the brain more slowly. As a result, many people experience a slowing in thought process and memory as they age.

Slowing of the thought process is not the same as senility, dementia, or severe memory loss. These conditions are not symptoms of the natural aging process. Older people who are showing signs of senility, dementia, or serious memory loss should consult a physician. These symptoms could indicate a more serious physical problem, such as Alzheimer's disease.

ALZHEIMER'S DISEASE

Alzheimer's disease is a medical condition that causes the brain to lose more and more nerve cells. The condition is one of the chief causes of dementia. Dementia is defined as changes in two or more of the following areas: language skills, decision-making abilities, judgment, attention span, and personality. Problems with these functions become worse as time goes on.

Alzheimer's is a common condition in the United States. About 4 million Americans suffer from the disease. Experts expect this number to increase steadily over the next few decades as American life expectancy continues to rise. Doctors are not sure what causes the condition. A diagnosis of Alzheimer's is made only after all other causes of dementia are ruled out. The diagnosis can be positively confirmed only after death, when brain tissue can be examined.

Some early warning signs of Alzheimer's include saying things repeatedly, frequently losing objects, difficulty naming common objects, becoming lost on familiar routes, and losing the enjoyment of hobbies and pastimes. More serious symptoms include forgetting life events, hallucinations, violent behavior, agitation, and depression. As the disease progresses, patients lose the ability to care for themselves. As brain cells die, such abilities as understanding language and identifying family members are lost forever.

People most at risk for Alzheimer's are the elderly, especially those with a family history of the condition. According to the Alzheimer's Association, as many as 50 percent of all people over the age of eighty-five have the disease. Alzheimer's may progress slowly or quickly. The average duration of the disease is about eight years, but a person with Alzheimer's may live as long as twenty years after the onset of symptoms.

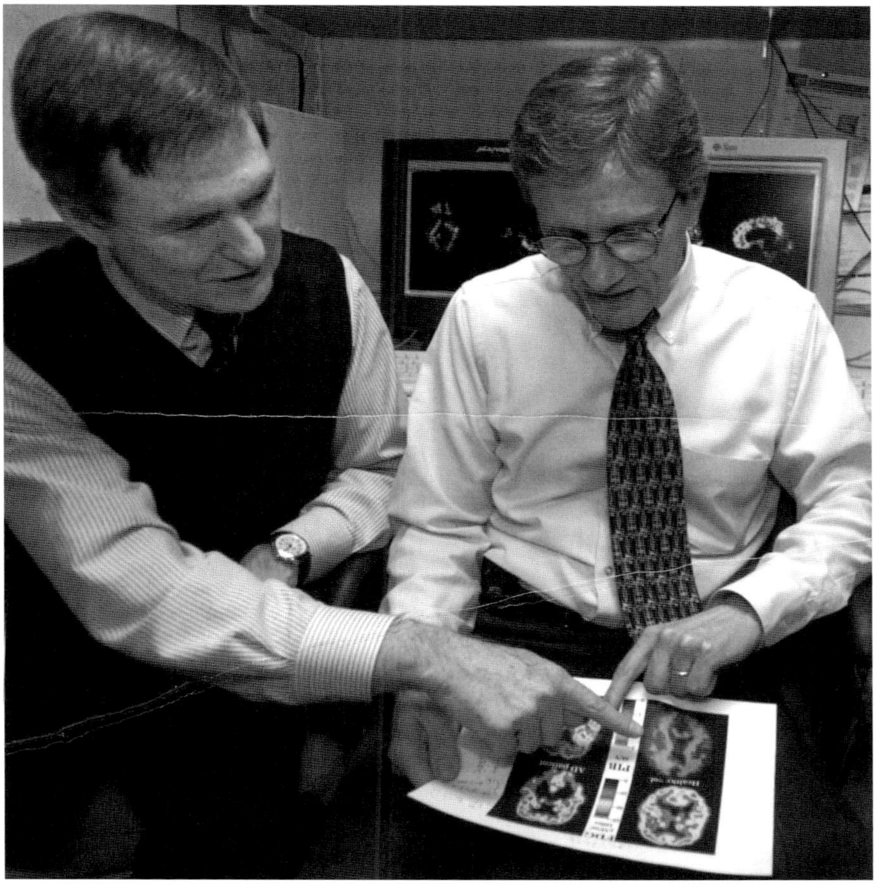

Psychiatrist William Klunk (right) and Chester Mathis (left), a radiochemist, look over the differences between the brain scans of a person affected by Alzheimer's disease and that of a healthy person. New technology is opening a window on the living human brain that allows researchers and physicians to see the effects of Alzheimer's disease as they happen.

There is currently no cure for Alzheimer's disease. Doctors use drugs to slow the progression of symptoms and improve memory and thinking skills in patients. Recent research has shown that keeping an active mind may stave off the disease. Activities that may help prevent the condition include reading, playing cards and other games, and listening to the radio.

AMYOTROPHIC LATERAL SCLEROSIS

Sometimes known as Lou Gehrig's disease for the famous baseball player who contracted it, *amyotrophic lateral sclerosis* (uh-mye-uh-TRAH-fik LAT-er-ul skluh-ROH-siss), or ALS, is a disease in which nerve cells in the spine and brain that control muscle function are destroyed over time. The condition affects about one in 100,000 people. The first nerve cells to be affected are those that control voluntary muscle function, such as arm and leg movement. Eventually, nerves controlling involuntary functions, such as breathing and swallowing, may also be affected. However, the parts of the brain that control thinking are not damaged.

Physicist Stephen Hawking (left) looks at a new custom-built computer designed especially for him by the Intel Corporation, a maker of computer chips and components. Hawking, who is considered one of the most brilliant scientists in history, has suffered from amyotrophic lateral sclerosis since his twenties. He uses the special computer to communicate.

The first symptoms of ALS usually appear after a person has reached the age of fifty. These symptoms include muscle weakness that becomes progressively worse; muscle cramps; paralysis; changes in the voice, hoarseness, or difficulty speaking; difficulty swallowing or breathing; and drooping head due to weakening spinal and neck muscles. Eventually, patients lose the ability to care for themselves. Death usually results between two to ten years from the onset of symptoms, often from respiratory failure.

Doctors are not sure what causes ALS, although it appears to run in some families. There is no known way to prevent or cure this condition, although some medicines can slow the onset of symptoms and ease symptoms once they do appear. Physical therapy may also slow the progression of the disease.

AUTISM

Autism is a developmental disorder of the brain. The condition, which is usually diagnosed during the first three years of life, affects a child's social and communication abilities. According to the Autism Society of America, 1.5 million Americans are affected by autism. Doctors are not sure what causes the condition, but genetic factors seem to play a role. Boys are affected by autism three to four times more frequently than girls are.

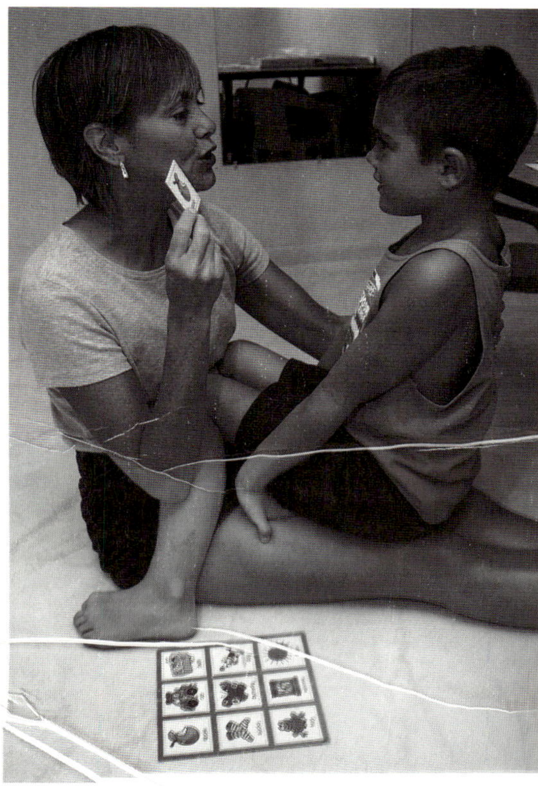

Ryan Coleman, who was born with autism, is home-schooled by his mother in a word association game that helps with reading and verbal skills.

Symptoms of autism include communication difficulties, short attention span, difficulties with social interaction, and repetitive body movements. About three out of ten people with autism also experience seizures. Cases of autism can range from mild to severe. Some children with mild autism, for example, may grow up to live on their own and fully function in society. Those with severe autism, however, may need constant care for the rest of their lives. Most autistic adults will need some assistance in their daily lives.

There is no cure for autism. Doctors treat the condition with therapies that address each individual child's problems. These treatments include physical and speech therapy, music classes, and vision therapy. Medications may also be prescribed.

BELL'S PALSY

Bell's palsy is facial freezing caused by damage to one of the cranial nerves. The cranial nerves are twelve sets of nerves that extend directly from the brain. Most cranial nerves transmit and receive motor or sensory information. Bell's palsy begins suddenly and unexpectedly and gradually becomes worse. Researchers are unsure of what causes the damage to a cranial nerve. They do know that many people with Lyme disease, an infectious bacterial illness transmitted by ticks, also suffer from Bell's palsy. Head injuries, tumors, and other medical conditions may also be to blame.

Symptoms of Bell's palsy include pain behind or in front of the ear, loss of the sense of taste, headaches, stiffness and feelings of pulling in the face, difficulty eating, drooping of the facial features, drooling, and paralysis on one side of the face. Doctors treat the condition with medicines that will reduce the swelling or pressure on the affected nerve. Between six and eight out of every ten patients suffering from Bell's palsy will recover within a few weeks or months. In some cases, however, a person may suffer permanent disfiguration, spasms, or eye damage.

BRAIN TUMORS

Brain tumors are abnormal growths of tissue in or on the brain. Brain tumors may be either benign or malignant. Benign tumors usually stay in one spot and are slow-growing. However, these types of tumors can still cause problems for people. The tumors put pressure on the brain, affecting nervous-system functions and brain activity.

While benign tumors usually stay in one spot, malignant tumors are more likely to spread to other sections of the brain. Some malignant brain tumors begin growing in the brain itself. These types of tumors are known as *primary tumors.* Most brain tumors in children are primary tumors. Sometimes, cancer cells migrate from other parts of the body to the brain. Tumors that form from cancer cells from areas outside the brain are known as *secondary tumors.*

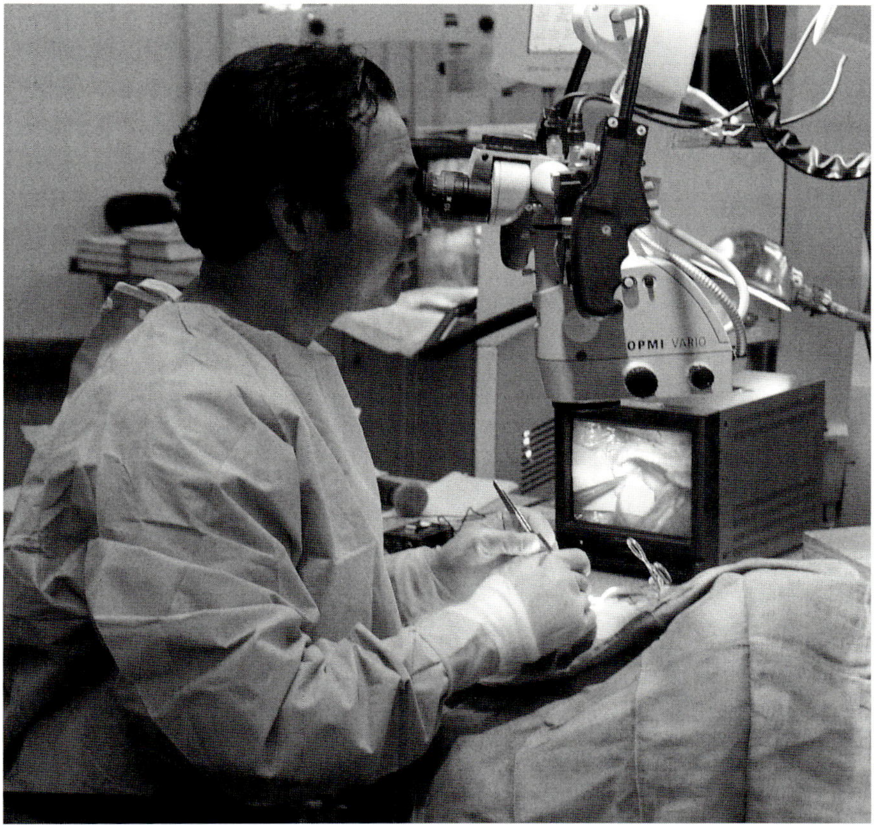

Neurosurgeon Saleem Il Abdulrauf of Saint Louis University performs a new cerebral bypass procedure that offers hope for patients with previously inoperable brain tumors and aneurysms.

Common symptoms of brain tumors are seizures, coordination problems, headaches accompanied by nausea and vomiting, slurred speech, dizziness, and vision problems. Treatment of the tumor will vary, depending upon the tumor's location, size, and type. When possible, doctors perform surgery to remove all or part of the tumor. Radiation therapy can shrink or destroy tumor cells. Chemotherapy and steroid treatment are also used to treat brain tumors.

Treatment is not a guarantee of a cure, even when a brain tumor is totally removed. Tumors may grow back. There is no way to tell whether they will, so people with brain cancer must be examined regularly to check for new tumors.

Fast Fact

Brain tumors are the second most common type of childhood cancer, after leukemia.

CEREBRAL PALSY

Cerebral (suh-REE-brul) *palsy* is a congenital (present at birth) disease that affects a person's motor skills, muscle tone, and muscle movement. The condition is usually the result of a brain defect or damage that occurs before or during birth. Risk factors for the condition include lack of oxygen, premature birth, low birth weight, infection, and nervous-system problems.

Cerebral palsy is a common disorder in the United States today. Each year, 5 out of every 2,000 babies born in the United States have cerebral palsy. The condition most frequently affects premature babies. In fact, 5 percent of all children born prematurely suffer from cerebral palsy. The number of cases is rising as more and more premature babies are saved by improved medical treatment. The earlier an infant is born, the higher the chances that the baby will have the condition.

Todd Gatewood of Coshocton, Ohio, uses a device called the "Liberator," which enables him to synthesize words into sentences. Gatewood, who suffers from cerebral palsy, is unable to speak clearly or use his legs.

Symptoms of cerebral palsy include serious delays in an infant's development, coordination problems, seizures, bladder and bowel-control problems, breathing disorders, learning disabilities, loss of vision and hearing, and difficulty eating. The severity of cerebral palsy varies. Some people may experience only mild symptoms. Others may require wheelchairs, braces, or other implements to help them function. Still others may be completely unable to care for themselves.

There is no cure for cerebral palsy. Physical therapy, speech therapy, and other therapies can help patients improve and maintain body function. Medicines and surgery are also sometimes used, depending upon the type and severity of the cerebral palsy.

EPILEPSY

Epilepsy (EH-pih-lep-see), also known as a seizure disorder, is a brain disorder that results in frequent seizures. During a seizure, the electrical signals in the brain misfire. A seizure may be triggered by flashing lights, repeating sounds, illness, or hormonal changes. In many cases, no trigger for the seizure is ever determined.

Epileptic seizures affect about 2 million Americans every year. Most epileptics are either younger than fifteen or older than sixty-five. However, any person of any age can suffer a seizure. Seizures may be the result of other medical conditions, such as brain tumors, brain infections, autism, stroke, or kidney failure. Doctors diagnose epilepsy after all other health conditions have been ruled out.

Fast Fact

Not all seizures are caused by epilepsy. For example, some children may experience seizures when they have high fevers of 102°F (39°C) or higher.

There are two major types of seizures: generalized and partial. *Generalized seizures* affect all or most of the brain, while *partial seizures* affect one specific location. Most epileptic seizures are generalized seizures. Two major types of generalized seizures are known as petit mal and tonic-clonic seizures. Symptoms of a petit mal seizure include a sudden loss of awareness with little body movement. The person having the seizure may seem to stare off into space. Children most often suffer from this type of seizure, which can interfere with their ability to learn. Symptoms of a tonic-clonic seizure, also known as a grand mal seizure, include violent muscle contractions, loss of consciousness, stiffness, loss of bladder control, and confusion and weakness afterward.

Just before a seizure, some epileptic patients may experience what is known as an *aura.* An aura is the sensation that indicates that a seizure is about to begin. People suffering grand mal seizures must be monitored carefully to make sure that they do not harm themselves or stop breathing. During the seizure, bystanders should place the person on his or her side to prevent the swallowing of vomit or mucus, but should not try to restrain the person or place anything in the person's mouth. A person who has suffered a grand mal seizure should seek medical attention. If the seizure lasts for longer than two or three minutes, emergency medical attention should be sought immediately. The person could suffer permanent brain damage from lack of oxygen to the brain.

Doctors treat epilepsy with medicines that prevent seizures. Although epilepsy is usually a lifelong condition, some children do outgrow the disorder.

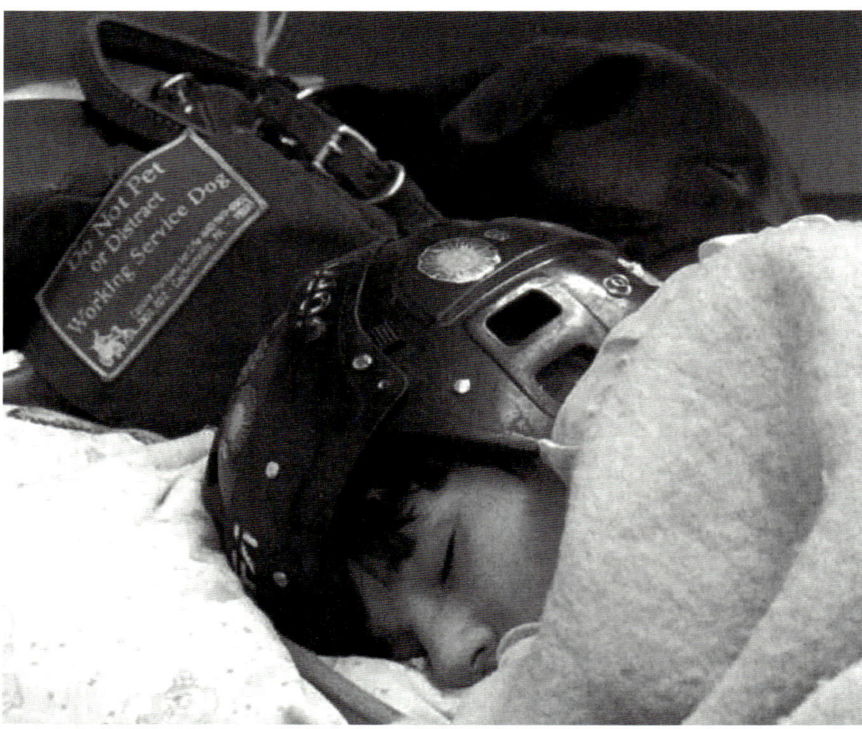

Emily Ramsey's dog, Watson, stays near her while she rests after having a seizure at Starbuck Middle School in Racine, Wisconsin. Watson is able to detect Emily's seizures before they are apparent to humans and help Emily communicate the onset of a seizure to her teachers so it can be effective y dealt with.

GUILLAIN-BARRÉ SYNDROME

Guillain-Barré (gee-LAHN buh-RAY) *syndrome* is a disease that attacks part of the nervous system, causing the nerves to become inflamed and damaged. As a result, muscles become weaker. Some people may even suffer paralysis. Doctors are not sure what causes the condition. They do know that people between the ages of thirty and fifty are more likely to get Guillain-Barré. Researchers have also found that the condition often appears after a patient has suffered a minor viral or bacterial infection.

Symptoms of Guillain-Barré include weakness, usually beginning in the legs, followed by paralysis, numbness or pain in the muscles, blurred vision, and clumsiness or falling. The condition often worsens quickly, and people may need to be hospitalized so that they can be helped to breathe. Patients with Guillain-Barré usually recover after a period of months. People who have suffered severe cases may need physical therapy to recover their muscle function.

HUNTINGTON'S DISEASE

Huntington's disease is a condition in which the nerve cells in a certain area of the brain waste away, causing the body to twist and writhe uncontrollably. It is an inherited condition that results from an abnormality in one gene. Any child who has a parent with Huntington's disease has a 50 percent chance of inheriting the disease.

Symptoms of Huntington's disease include such behavior changes as irritability, restlessness, paranoia, hallucinations, and dementia. Other changes include uncontrollable facial movements, an unsteady gait, sudden jerking of the body, speech problems, and difficulty swallowing. There is currently no cure for Huntington's disease. Doctors treat the condition with medication aimed at slowing the disease's progression. Eventually, however, people with Huntington's disease lose the ability to care for themselves.

MULTIPLE SCLEROSIS

Multiple sclerosis (MS) is a disease of the central nervous system (brain and spinal cord). The protective covering on nerves of the central nervous system becomes damaged and destroyed, leaving behind scar

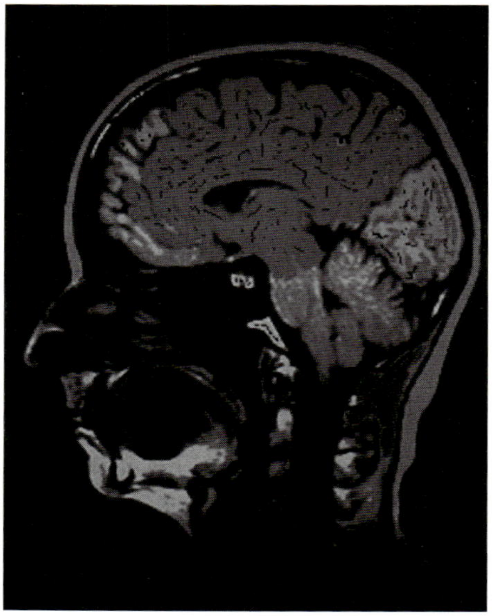

This MRI (magnetic resonance image) shows the brain of a person afflicted with multiple sclerosis. Doctors use MRIs to detect and diagnose MS.

tissue. Scar tissue slows or blocks nerve messages into and out of the central nervous system. About one in every 1,000 Americans is affected by MS. Women are more commonly affected than are men.

Doctors do not know what causes MS. Some believe that it is an autoimmune condition. MS can affect patients in many different ways, depending upon which nerves are damaged. Some patients suffer few symptoms. Others become completely disabled by the disease. Symptoms of MS include weakness; paralysis; tremors or pain in the extremities; muscle spasms; atrophy (AA-truh-fee), or wasting of the muscles; numbness, tingling, vision problems, loss of balance and coordination, vertigo (dizziness), urination problems, difficulty speaking, fatigue, decreased memory, judgment, and attention span; and depression. Symptoms may last for days or months. Then they may disappear, only to recur again.

There is currently no cure for MS, and it is impossible for doctors to predict how quickly the condition will progress in each patient. Doctors treat the disease by controlling the symptoms and keeping the body as healthy as possible. Treatment may include the use of medicines, physical therapy, speech therapy, and exercise.

PARKINSON'S DISEASE

Parkinson's disease is a degenerative brain disorder. Nerve cells within the part of the brain that controls muscle movement begin to deteriorate. As a result, movement and coordination are greatly affected. Slowly, the disease takes away a person's ability to function on a day-to-day basis.

Parkinson's disease affects two out of every 1,000 Americans. Those most at risk are over the age of fifty, although children and younger adults may also be affected by the disease. Parkinson's is the most common neurological disease of the elderly. Doctors are not sure what causes most cases. However, research suggests that Parkinson's in younger people may be caused by genetic factors.

Symptoms of Parkinson's include tremors, rigid muscles, slowness of movement, slumped posture, loss of balance and coordination, muscle aches, loss of facial expression, voice changes, decline in mental function, depression, dementia, hallucinations, memory loss, and drooling. Doctors determine if a person has Parkinson's based on these symptoms and after ruling out other disorders.

There is at present no cure for Parkinson's disease. Doctors do their best to treat the many symptoms associated with the disorder. A number of medications, for example, can be used to increase dopamine (DOH-puh-meen) production in the brain. Dopamine is a chemical that is essential to the proper functioning of the central nervous system. Doctors also recommend that patients exercise, eat a healthy diet, and stay as active as possible. Less common treatments include surgery to destroy tremor-causing tissue. If left untreated, Parkinson's disease can become disabling and may cause early death.

DIAGNOSING AND TREATING NERVOUS SYSTEM DISORDERS

Doctors have a number of tools at their command to help them diagnose nervous-system disorders. For example, doctors can use a number of different imaging tests to look inside the human body. CT scans and MRIs are the most commonly used tests to take pictures of the brain. *Sonograms,* using sounds waves, can also be used to learn more about a person's body. *Arteriograms* (ar-TEER-ee-oh-gramz) take pictures of the arteries that run to and from the brain.

One imaging test that is of particular importance when mapping the brain is *positron emission tomography* (PET). First, a radioactive substance is injected into the body. Then pictures are taken as the substance moves through the body. PETs allow doctors to learn more about Alzheimer's and Parkinson's diseases.

Electrodiagnostic (ee-lek-troh-dye-ag-NAH-stik) tests are used to check the functioning of motor neurons and muscle reactions. The tests use electrodes, placed over or into the muscles, to check muscle response. One electrode test, called an *electroencephalogram* (ee-lek-troh-en-SEF-uh-luh-gram), or EEG, specifically measures brain activity by using electrodes attached to the scalp.

Doctors perform a *lumbar puncture,* also known as a spinal tap, by withdrawing cerebrospinal fluid from the spine with a needle. The fluid is then tested for signs of infection.

Doctors also have a variety of methods to use when treating certain nervous-system disorders. For example, in a *cryothalamotomy* (krye-oh-thal-uh-MAH-tuh-mee), a supercold probe is inserted into the thalamus in order to stop tremors. Another treatment is a *pallidotomy* (pal-ih-DAH-tuh-mee). During this procedure, a section of the brain is destroyed to stop tremors and other symptoms associated with Parkinson's disease.

Surgical resection is another treatment for nervous disorders. With this method, all or part of a brain tumor is removed through surgery. Other options for treatment include chemotherapy and radiation therapy.

9

HOW THE ENVIRONMENT AFFECTS THE NERVOUS SYSTEM

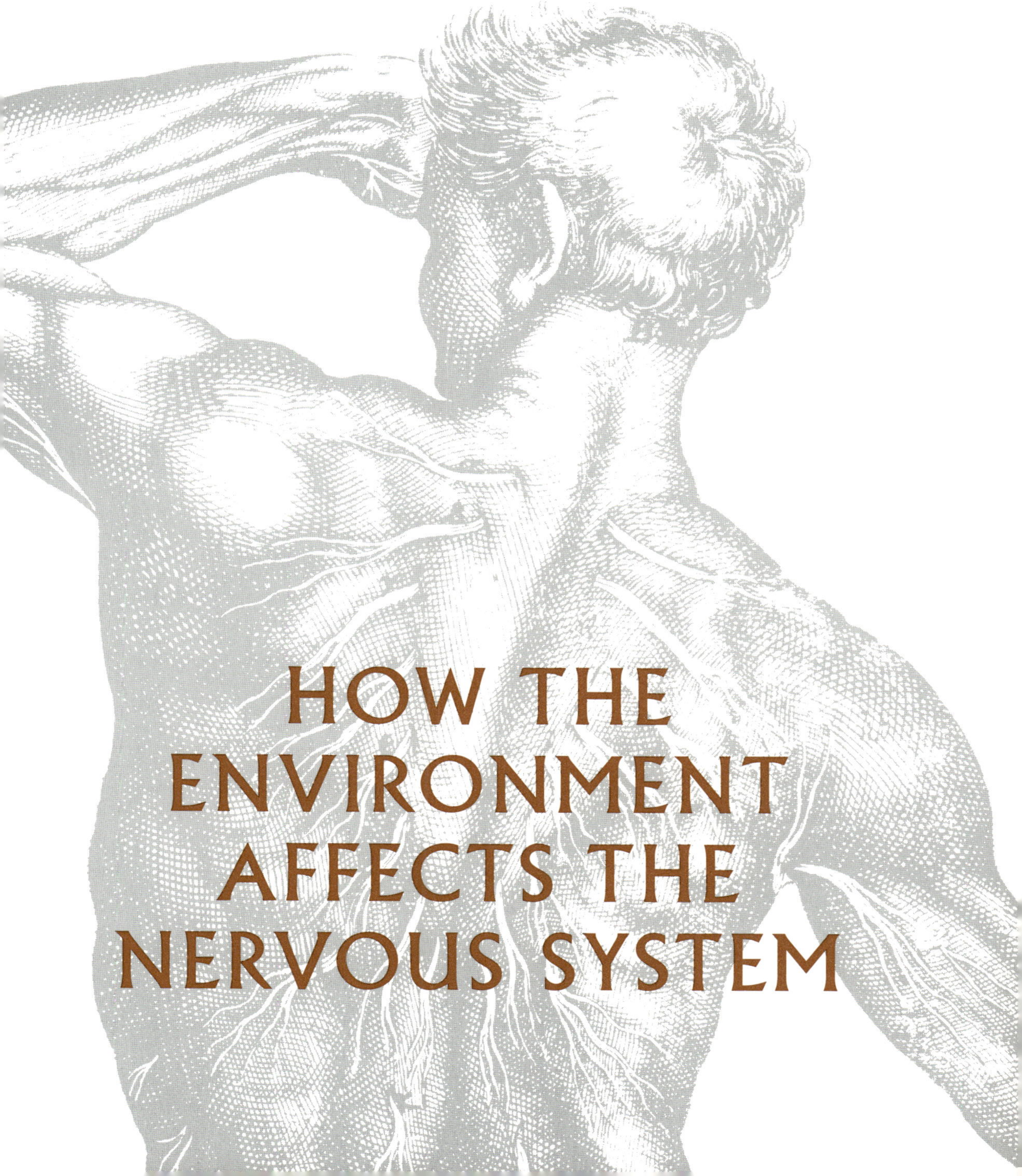

The environment can seriously affect the nervous system. The use of alcohol and drugs, a poor diet, and trauma and injury can damage the nervous system. Even stress can cause or worsen medical conditions for people.

ALCOHOL AND OTHER DRUGS

Alcohol can be extremely damaging to the nervous system. Mild to moderate use of this drug can cause a person to lose some ability to learn and remember. In addition, the brain can be seriously and permanently damaged as a result of long-term alcohol abuse. People who drink too much alcohol at one time may engage in risky behaviors, such as choosing to drive while drunk. This may lead to traumatic injuries as a result of accidents.

One nervous-system condition associated with alcohol abuse is *alcoholic neuropathy* (noor-AH-puth-ee). With this disorder, nerves become damaged as a result of excessive alcohol use. Early symptoms of alcoholic neuropathy include numbness, tingling, muscle cramps and weakness, constipation or diarrhea, and nausea and vomiting. Symptoms of more serious cases include difficulty swallowing, moving, and talking, muscle atrophy, hoarseness, and drooping eyelids. There is no cure for this condition. Doctors can treat the symptoms of the disorder, but there is usually no way to reverse existing damage to the nerves. Further nerve damage can occur if the patient refuses to give up alcohol.

Another medical condition caused by excessive alcohol intake is *dry beriberi* (berr-ee-BERR-ee). Dry beriberi results from a deficiency of vitamin B1, or thiamine. This deficiency is most often the result of alcohol abuse. People who abuse alcohol often do not have healthy diets. However, dry beriberi can also affect people in developing countries who have restricted diets; people undergoing dialysis (dye-AL-uh-siss), a medical procedure to remove wastes and toxins from the blood; and people taking large doses of certain types of medicines. Symptoms of the disorder include pain, tingling, or loss of sensation in the feet and hands. Serious cases can result in brain damage, coma, and death. Patients are usually treated with thiamine, but the most serious symptoms may be impossible to treat.

Wernicke-Korsakoff (VER-nih-kee KOR-suh-koff) *syndrome* is yet another disorder that can be caused by alcohol abuse. Like dry beriberi, the syndrome is often the result of a deficiency of thiamine. The condition occurs when nerves in the central and peripheral nervous systems become damaged. Symptoms include vision changes, loss of muscle coordination, loss of memory, the inability to store new memory, and hallucinations. Doctors treat the condition by hospitalizing patients in order to control the symptoms and make sure the illness does not get any worse. Thiamine supplements may also be given to prevent further damage.

Statistics show that people who are addicted to smoking are often also addicted to alcohol. Some researchers believe that giving up smoking could help people who are battling alcoholism.

Alcohol poisoning can affect anyone who drinks too much alcohol at one time. Alcohol depresses the central nervous system, slowing nerve functions that control such involuntary actions as breathing and the gag reflex. A person suffering from alcohol poisoning may lose consciousness, choke on his or her own vomit, and die. Symptoms of alcohol poisoning include confusion, vomiting, seizures, slow or irregular breathing, hypothermia (hye-poh-THUR-mee-uh), or low body temperature, and coma. Death may result from choking or heart failure, and permanent brain damage is a possibility if the person loses consciousness for too long. Alcohol poisoning is a medical emergency that requires immediate professional attention.

Illegal drugs can seriously damage the nervous system. One of the most toxic is cocaine. Cocaine, a stimulant, can cause brain infections and psychiatric disorders. It can also cause coma and death. The use of needles to inject illegal drugs can lead to bacterial infections of the spinal cord and the brain. Some people may also become addicted, or mentally and physically dependent on illegal drugs.

In recent years, some criminals have sexually assaulted women after secretly putting illegal drugs in the women's drinks. These drugs, sometimes called "date rape drugs," are central nervous-system depressants that can be easily mixed into drinks at clubs or bars. They cause unconsciousness, allowing a victim to be easily assaulted. These drugs can also cause coma and death.

ACCIDENTS AND TRAUMA

Accidents that damage the brain or spinal cord can permanently alter a person's life. Such traumas can impair a person's ability to function on a day-to-day basis and even bring about personality changes.

TRAUMATIC BRAIN INJURY

Traumatic brain injury (TBI) is a condition in which a person suffers any type of sudden damage to the brain. The most common causes of TBI are motor-vehicle accidents, firearm incidents, falls, workplace accidents, recreational accidents, assaults, and accidents in the home. TBI is a common injury in the United States. Each year, about 270,000 people suffer a moderate or severe case of TBI. More than 53,000 Americans die as a result of TBI each year. Those most commonly affected by TBI include young males between the ages of fifteen and twenty-four, elderly people over the age of seventy-five, and children under the age of five.

TBI can range in seriousness from mild to severe. Mild cases of TBI can cause headaches or a loss of consciousness. Severe cases can cause a loss of mental ability, permanent personality changes, and death. The injuries associated with TBI include *contusion,* or a bruising of the brain; *shearing,* or damage to the brain's nerve cells; and *hematoma* (heem-uh-TOH-muh), or heavy bleeding inside or around the brain. *Anoxic brain injury* results in a lack of oxygen to the brain, causing brain cells to die.

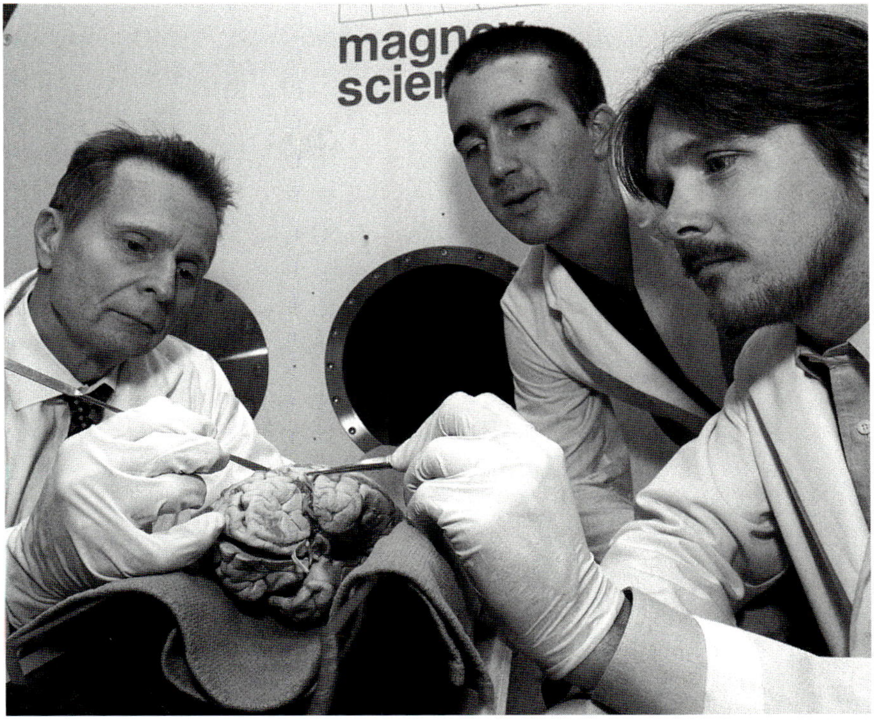

Three scientists prepare a preserved human brain for anatomical imaging at the University of Florida's Brain Institute and Center for Traumatic Brain Injury (TBI) Studies. The scientists hope that by studying brains affected by TBI, they can better understand how to help people with brain injuries.

The symptoms of TBI may appear immediately after a head injury or develop slowly, over a period of hours. Symptoms of mild TBI include nausea, dizziness, fatigue, loss of consciousness, headaches, blurred vision, ringing in the ears, and trouble thinking, concentrating, or remembering. These symptoms may last for several days. Symptoms of more serious cases of TBI include more severe and long-lasting headaches, nausea and vomiting, seizures, inability to awake after falling asleep, dilation of one or both pupils, slurred speech, numbness of the arms and legs, confusion, and loss of coordination. These symptoms indicate a medical emergency, and a patient should seek medical attention immediately.

Coma is one of the symptoms of serious TBI. Coma is a condition in which a patient is completely unconscious. Patients are unaware of their surroundings and cannot be awakened manually. Most comas last from a few days to a few weeks. Some coma patients eventually improve and come out of comas on their own. Others may worsen and die.

Another result of TBI is a *vegetative state.* Patients in a vegetative state, like those in a coma, are unconscious and unaware of their surroundings. However, these patients can open their eyes and may move or groan at times. They have sleeping and waking moments and periods of alertness. About half of all adults in a vegetative state recover within the first six months. Experts say that patients who spend more than a year in a vegetative state are unlikely to ever recover.

Brain death occurs when the brain has been so badly damaged that it cannot continue to function. All activity in the brain and the brain stem ceases. The body of a patient who suffers brain death can be kept functioning only with lung and heart machines. The family of a patient who has suffered brain death may choose to donate his or her organs. These organs will be transplanted into people who are in need of healthy organs.

After a person suffers a TBI, doctors use X-rays and CT scans to determine the amount of damage to the brain. Because it is impossible to repair damage that has already been done, the goal of treatment is to stabilize the patient's condition and prevent further damage. In cases of mild TBI, no treatment may be needed, and the patient recovers over time. For more serious injuries, doctors may stabilize the patient by surgically draining fluids from the brain to relieve pressure. Doctors may also prescribe medicines that relieve brain swelling.

TYPE OF TBI	Concussion
NATURE	The brain has been violently shaken.
SEVERITY	Minor to moderate
TYPE OF TBI	Depressed skull fracture
NATURE	Pieces of fractured skull are crushed inward and press against the brain.
SEVERITY	Moderate to severe
TYPE OF TBI	Penetrating skull fracture
NATURE	An object penetrates through the skull and into the brain.
SEVERITY	Moderate to severe
TYPE OF TBI	Shaken baby syndrome
NATURE	An infant's brain bounces against the skull as a result of shaking.
SEVERITY	Severe

Avoiding TBI

TBI is usually the result of accidents. You can help prevent TBI by following these tips, recommended by the CDC.

- Always wear a seatbelt when riding in a motor vehicle. Encourage your friends and family to do the same.
- Always wear a safety helmet when cycling, skateboarding, rollerblading, horseback riding, or engaging in other activities where a fall is possible.
- Use the handrails on stairways to avoid a fall.
- When using a ladder or step stool, keep a firm grip on it.

Once the patient has been stabilized, doctors will begin to assess how seriously a person has been affected by the TBI. Treatment will depend upon which parts of the body remain affected by the injury. For example, if a patient has forgotten how to walk as a result of a TBI, physical therapy will be part of the treatment. For someone whose speech areas of the brain are affected, speech therapy will be recommended.

SPINAL-CORD INJURY

Injuries to the spinal cord or the bones and tissues that surround the spinal cord can disrupt the connection between the brain and the other parts of the body. A spinal-cord injury can result in serious damage at and below the site of injury, including loss of sensation and paralysis in parts or all of the body. In some cases, the injury can result in death.

Each year, as many as 15,000 Americans suffer spinal-cord injuries. About 10,000 of these victims are left permanently paralyzed. Many of the others die. Most sufferers of spinal-cord injuries are young men between the ages of fifteen and thirty-five. Common causes of spinal-cord injuries include bullet or stab wounds, motor-vehicle accidents, diving accidents, electric shocks, and sports injuries.

Symptoms of a spinal-cord injury include vomiting, a stiff neck, weakness, difficulty or inability to move or walk, shock, loss of bladder control, and loss of consciousness. If you suspect that a person has suffered a spinal-cord injury, it is important that the

Vicky Allen has overcome incredible odds following an accident that left her paralyzed. She and her husband had two children after doctors told her that the spinal-cord injury would prevent her from successfully carrying a baby.

person not be moved. Moving the patient could cause further, permanent damage. Call for emergency medical assistance immediately. The sooner a patient receives medical attention, the better the chances are for successful treatment.

To treat a spinal-cord injury, doctors use medications that reduce swelling around the injured area. In addition, surgery may be performed to remove fluid, tissue, or pieces of bone. A patient with a dislocated spine may need to be placed in traction. Once spinal-cord nerves have been damaged, they cannot be repaired. After patients have been stabilized and begin to recover, physical therapy may be necessary to help them adjust to any permanent damage.

Whiplash

Whiplash is a mild spinal-cord injury. It results from damage to the soft tissues of the neck as a result of a sudden, violent, backward motion of the head. This type of injury most commonly occurs as the result of motor-vehicle accidents. Whiplash symptoms, which may not be noticed until hours or even weeks after the injury, include dizziness, headaches, and pain in the neck, shoulders, jaw, and arms. Whiplash is usually treated with rest, painkillers, and a neck brace to keep the neck from moving. Most people recover within three months.

BACTERIAL AND VIRAL CONDITIONS

Viruses, bacteria, and other organisms can cause infection, inflammation, and injury to the nervous system. Nervous-system infections include encephalitis, meningitis, and polio.

ENCEPHALITIS

Encephalitis (en-sef-uh-LYE-tiss) is an inflammation of the brain most often caused by a virus. As the brain swells from the infection, the swelling can destroy nerve cells or cause bleeding within the brain and permanent brain damage. In the most serious cases, encephalitis can lead to death.

Many different types of viruses, as well as some bacteria, can cause encephalitis. A person can be exposed to this rare disorder from insect or tick bites or contaminated food, or by inhaling secretions from other people. Encephalitis often follows another disease, such as chicken pox, measles, or mononucleosis (mahn-oh-noo-klee-OH-siss). (Mononucleosis is an infectious disease usually caused by a virus.)

Symptoms of encephalitis include fever, headache, nausea and vomiting, confusion, clumsiness, and sensitivity to light. More serious symptoms, which indicate that emergency care is needed, include seizures, loss of consciousness, coma, paralysis, and memory loss. In some cases, antiviral medications can be used to treat the disorder. In most cases, however, there is no specific treatment for the disease. Doctors recommend that patients rest and drink plenty of fluids. The worst of the disease symptoms usually last about one or two weeks, but recovery may take many months.

A crop duster plane sprays insecticides over Westerly, Rhode Island, in an effort to control the mosquito-borne virus that causes encephalitis.

MENINGITIS

Meningitis (men-in-JYE-tiss) is an inflammation of the membranes of the brain and/or spinal cord. The condition is most often caused by a viral, bacterial, or fungal infection. Most people with viral meningitis recover over time. However, bacterial meningitis can be much more serious, leading to brain damage and even death.

There are several different types of meningitis. The most common type is pneumococcal meningitis. It affects about 15,000 people in the United States each year.

TYPE OF MENINGITIS	Aseptic (ay-SEP-tik)
CAUSE	Virus or fungus
SYMPTOMS	Fever, headaches, nausea and vomiting, sore throat, stiff neck, muscle and abdominal pains, rash, drowsiness, sensitivity to light, and confusion
PROGNOSIS	Excellent; most people recover within two weeks
TYPE OF MENINGITIS	Cryptococcal (krip-toh -KAHK-ul)
CAUSE	Fungus
SYMPTOMS	Fever, headaches, nausea and vomiting, sensitivity to light, and hallucinations
PROGNOSIS	Varies, depending upon patient's preexisting health conditions
TYPE OF MENINGITIS	Meningococcal (men-in-goh-KAHK-ul)
CAUSE	Bacteria
SYMPTOMS	Fever, headaches, rash, sensitivity to light, stiff neck, and problems with mental function
PROGNOSIS	Serious, causing death in up to 15 percent of all patients
TYPE OF MENINGITIS	Pneumococcal (noo-moh-KAHK-ul)
CAUSE	Bacteria
SYMPTOMS	Fever, headaches, nausea and vomiting, sensitivity to light, stiff neck, and problems with mental function
PROGNOSIS	Serious, causing death in 20 percent of all patients and leaving 50 percent of all patients with long-term complications

POLIOMYELITIS

Poliomyelitis (poh-lee-oh-mye-uh-LYE-tiss), more commonly known as polio, is a viral infection caused by exposure to secretions from infected people. In some cases, the infection attacks the central nervous system, causing muscle paralysis and even death. This type of polio is called *clinical polio.*

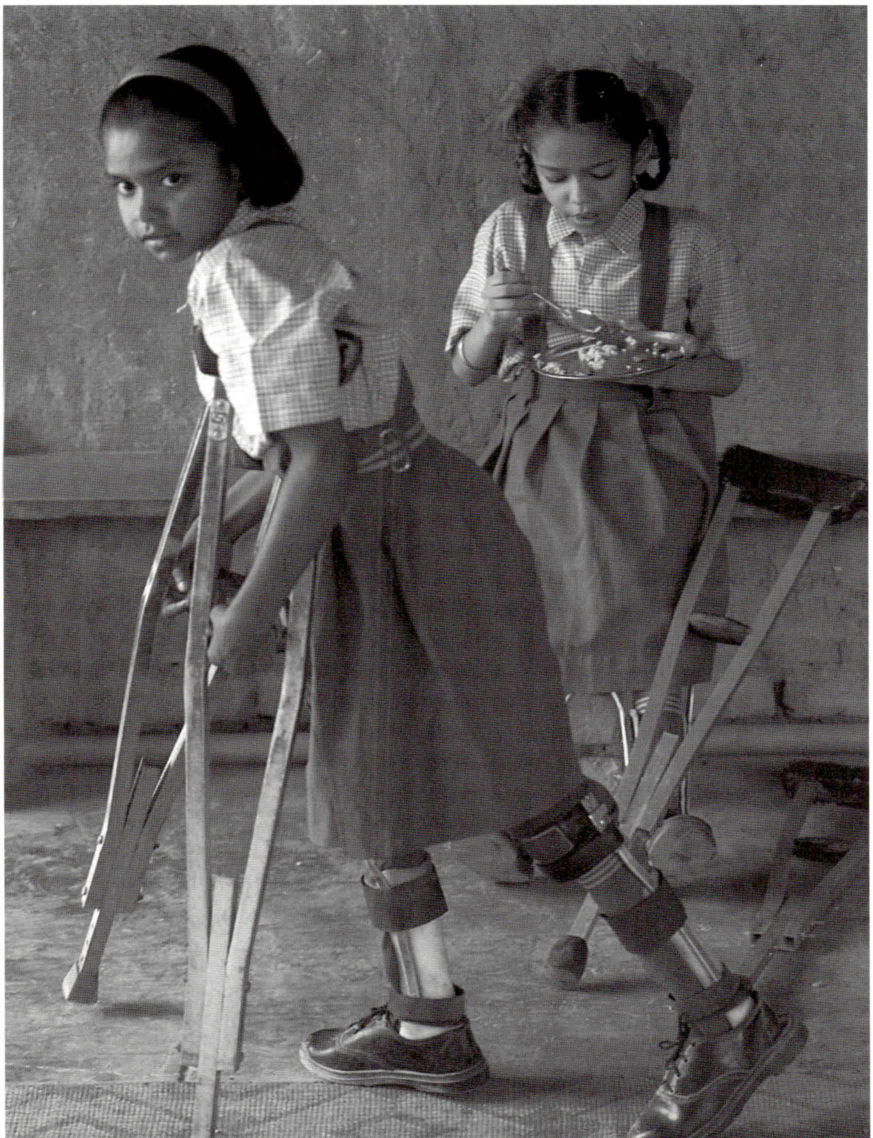

Children affected by polio have lunch at the Akshay Pratishthan School in New Delhi, India. Polio is still a health problem in parts of the world that have poor child health care.

Clinical polio is divided into two separate categories, nonparalytic and paralytic. Symptoms of *nonparalytic polio* include fever; headache, stiff neck, vomiting, diarrhea, neck, back, or leg pain; and stiffness, tenderness, or spasms of the muscles. Symptoms of *paralytic polio* include many of the symptoms of nonparalytic polio, as well as drooling, difficulty breathing, sensitivity to touch, and constipation. Doctors treat clinical polio by easing the symptoms and allowing the disease to run its course.

In 90 percent of all polio cases, damage to the spinal cord and brain can cause permanent injury, including paralysis and difficulty breathing. These long-term complications can, in turn, lead to other medical conditions, including heart problems, respiratory infections, high blood pressure, and urinary (YOOR-ih-neh-ree)-tract infections.

In the early twentieth century, polio was a common childhood scourge. In the late 1950s, however, a polio vaccine was developed by Dr. Jonas E. Salk (the Salk polio vaccine) that practically eliminated the disease in most industrialized countries. In some parts of the world, however, where people have poor medical care, polio still takes a deadly toll, especially on children. Half of all those who contract polio are under the age of three.

A United States President with Polio

Perhaps the most famous American to suffer from polio was Franklin Delano Roosevelt, the thirty-second president of the United States. Roosevelt contracted polio in the summer of 1921 while vacationing in Canada. Although the disease most commonly affects children, Roosevelt was thirty-nine when he developed it. The disease robbed him of the use of his legs, leaving him permanently paralyzed.

For the rest of his life, Roosevelt worked hard to hide his disability. He used a pair of heavy steel braces to enable him to stand upright, and with the help of an assistant and a cane, he could propel his hips forward to give the illusion of walking. During his twelve years as president, Roosevelt forbade reporters from ever photographing him while he was in his wheelchair. As a result, few Americans knew that their president was paralyzed.

10

THE SENSES

The senses allow people to fully experience the world around them. The body's five senses are hearing, sight, smell, taste, and touch. To make each sense possible, special organs work together with special nerve cells and the brain to interpret messages from the outside world.

HEARING

Hearing is the sense that is responsible for collecting and translating auditory messages from the outside world. The chief organ of hearing is the ear. The ear collects vibrations known as sound waves, turns them into nerve messages, and sends them on to the brain. Inside the brain, a message is translated as a particular sound. The ear is made up of three main sections: the outer, middle, and inner ear.

The function of the outer ear is to capture sound waves and direct them toward the middle ear. The outer ear is made up of the auricle and the ear canal. The *auricle* (AUR-ih-kul) is the flap of skin that is visible on the outside of the head. The auricle is specially shaped to catch sound waves and funnel them into the ear canal. The *ear canal,* a hollow tunnel about 1 inch (2.5 centimeters) long, connects the outer ear to the middle ear. It is lined with tiny hairs and sweat glands. These glands produce *cerumen* (seh-ROO-men), more commonly known as earwax. Earwax traps dirt and other foreign substances before they enter the middle ear.

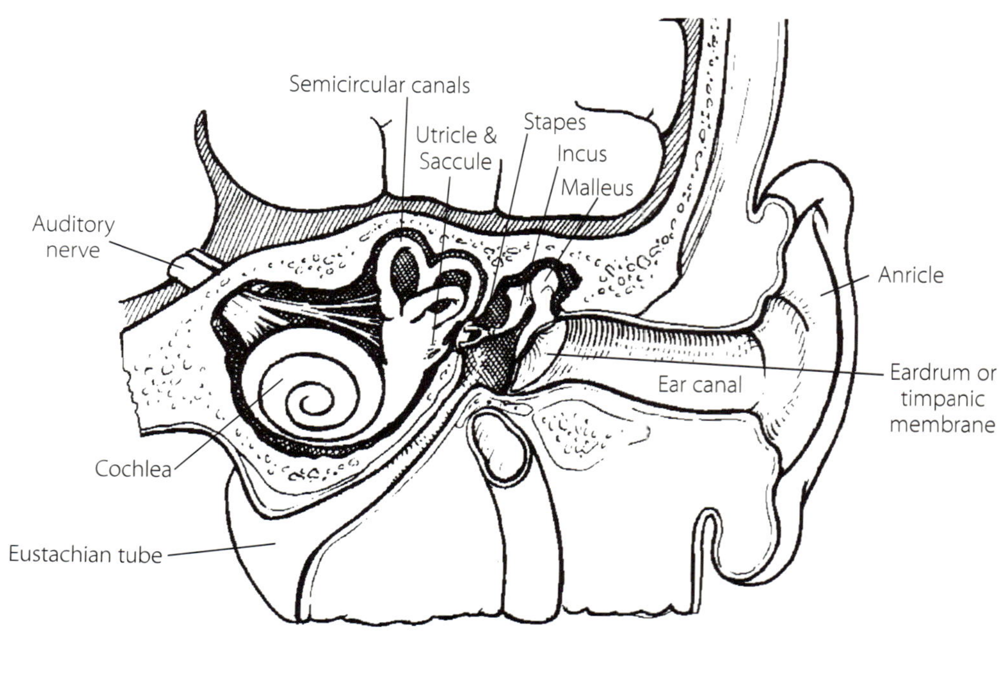

The middle ear is a hollow, air-filled chamber where sounds are amplified and passed on to the inner ear. The middle ear includes the *tympanic* (tim-PAN-ik) *membrane,* or eardrum; three tiny bones called the *malleus* (hammer), *incus* (anvil), and *stapes* (stirrup); and the *eustachian tubes.* Sound waves from the outer ear strike the eardrum, causing it to vibrate. This, in turn, causes the three tiny bones to vibrate. The vibrations are passed on to the inner ear. The eustachian tubes connect the middle ear to the back of the throat. They keep pressure equalized between the middle and outer ears.

Fast Fact

The hammer, anvil, and stirrup are the smallest bones in the human body.

The inner ear is where the vibrations are turned into nerve messages and sent along to the brain. The inner ear is made up of the cochlea, three semicircular canals, and two structures known as the *utricle* (YOO-trih-kul) and *saccule* (SAK-yool). The fluid-filled *cochlea* (KAHK-lee-uh), a coiled organ that resembles a snail's shell, is lined with thousands of nerve endings. As vibrations cause the fluid inside the cochlea to move, it stimulates the nerve endings. The nerves send messages to the brain, where they are translated as a particular sound.

SIGHT

Sight is the sense that is responsible for collecting and translating visual signals from the outside world. The eye is the organ that allows people to see the world around them. This seemingly simple organ, just 1 inch (2.5 centimeters) in diameter, is actually very complex. It is made up of a wall, a lens, and two inner chambers.

The wall of the eye is composed of three different layers: the sclera, choroid, and retina. The *sclera* (SKLUR-uh), more commonly known as the white of the eye, is the outermost layer. Only about one-sixth of the sclera is visible. In the center of the visible section of the sclera is the *cornea* (KOR-nee-uh). The cornea is a thin, transparent membrane that acts as a window, allowing light into the eye.

The *choroid* (KOR-oyd), below the sclera, contains blood vessels and dark pigment that absorbs light from the outside world. The visible section of the choroid is called the *iris.* People with brown eyes have choroid layers with considerable amounts of pigment, while those with blue eyes have less pigment. In the center of the iris is an opening called the *pupil.* The pupil can change size, becoming larger to let in more light or smaller to let in less light.

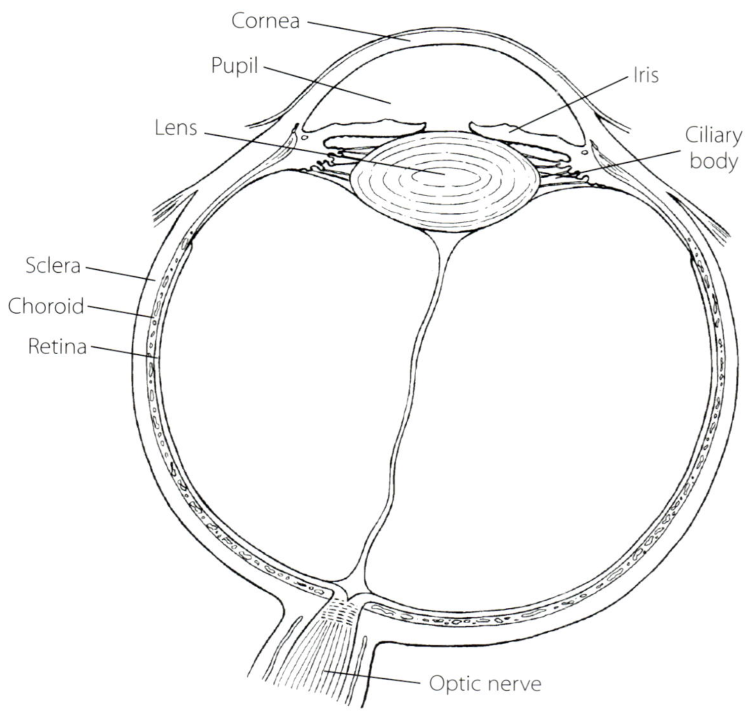

Behind the iris lies the *lens.* The oval-shaped lens is a clear, elastic structure that focuses light upon the retina. Muscles in the choroid layer, called the *ciliary* (SIL-ee-err-ee) *body,* cause the lens to become fatter or thinner. The lens changes shape depending upon whether an object is nearby or far away.

The *retina* (RET-ih-nuh) is the innermost layer of the eye wall. The retina contains millions of special cells called *cones* and *rods.* These cells convert light to messages that are sent to the brain. Cones translate color, bright images, and images that are directly in front of a person. Rods translate dim images, as well as peripheral, or side, images.

The two hollow inner chambers of the eye are separated from one another by the lens. The small cavity in front of the lens is filled with a watery liquid called *aqueous* (AK-wee-uss) *humor.* Aqueous humor helps the eye keep its shape. It also supplies important nutrients to the lens and cornea. The larger cavity behind the lens is filled with a gel-like substance called *vitreous* (VIH-tree-uss) *humor.* Vitreous humor also helps maintain the eye's shape.

The process of seeing begins when light waves enter the eye. Light passes through the cornea, pupil, and lens. The lens bends the light rays, focusing them on the retina to produce a clear image there. When the light rays hit the retina, cones and rods convert them into nerve messages to be sent to the brain. When the messages reach the brain, they are translated as images. The entire process of seeing something takes merely a thousandth of a second.

SMELL

The sense of *smell,* also known as olfaction, is the ability to detect odors by breathing in scent molecules in the air. The nose is the organ that controls the sense of smell. High inside the nose is a space called the olfactory area. This area, made up of a thin layer of mucus, contains special olfactory cells. The olfactory cells detect odor molecules from food, flowers, perfumes, and all the other "smelly" things in the world around us. Once the cells detect these molecules, they send messages to the brain. The brain interprets the messages as a particular scent.

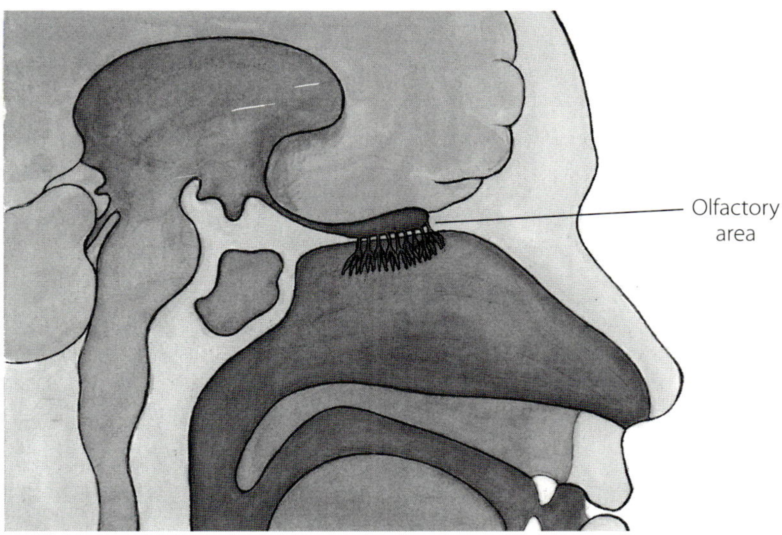

The olfactory membrane is shown in this illustration.

Fascinating Facts about Smell

- Experts say that the sense of smell is most accurate between the ages of thirty and sixty.
- Women have a more accurate sense of smell than men do.
- Experts say that there are a limited number of odors that combine to form a variety of different scents. Estimates of the exact number of odors range from seven to more than fifty.
- Humans can detect about 10,000 different scents.

TASTE

Taste, also called gustation, is the ability to detect the molecules of flavors from food and other substances placed into the mouth. The chief organ of tasting is the *tongue.* The tongue is covered with thousands of clusters of small organs called *taste buds.* Taste buds identify four different flavors: sweet, salty, sour, and bitter. Taste buds on certain sections of the tongue are better at identifying certain flavors. For example, the taste buds on the front of the tongue are better at detecting sweet flavors than other taste buds are, while those on the sides are better at sensing sour and salty flavors. The back of the tongue is where bitter flavors are identified.

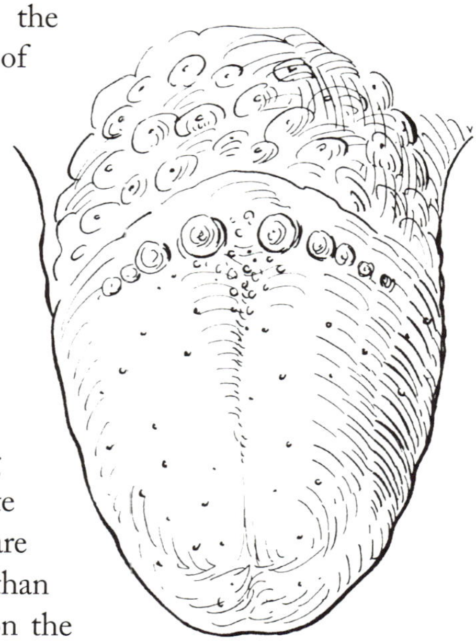

This is a drawing of the tongue.

Taste begins when a person puts a piece of food into the mouth. As the person chews, food is broken down and mixed with saliva (suh-LYE-vuh). Saliva carries the molecules over the tongue. Special cells within the taste buds, called *gustatory cells,* convert the molecules into nerve messages. The messages are then sent to the brain, where they are translated as a specific flavor or combination of flavors.

Nerve cells in the mouth also send information about the food's temperature and texture. Along with smell information from the nose, these messages provide a complete sense of how a food tastes.

Fascinating Facts about Taste

- People have nearly 10,000 taste buds.
- Most taste buds are located on the surface of the tongue. However, some taste buds are found in the throat and on the inside of the mouth.
- People are most sensitive to bitter tastes.
- Although the tongue can identify four main flavors, combinations of these flavors allow people to enjoy many different tastes.

Taste and Smell: What's the Connection?

Flavor comes from a combination of taste and smell. The enjoyment of eating a piece of chocolate cake comes partly from the smell of the chocolate and partly from the taste. A person who has lost the sense of smell may also experience a decline in the sense of taste. As a result, those who have permanently lost their sense of smell often eat highly spiced meals in order to enjoy taste sensations.

It is easy to test the connection between the sense of taste and smell. The next time you eat a piece of candy or something else that you normally enjoy, hold your nose. You will notice that the flavor doesn't seem quite as delicious as it does when you can also smell the food.

TOUCH

Touch is the ability to feel physical sensations on the surface of the body. These sensations include pressure, heat, cold, pain, roughness, and smoothness. The skin, the largest organ in the human body, allows people to experience these physical sensations. The skin has other important functions, as well. It covers and protects the entire outer surface of the human body from contaminants. It also helps regulate body temperature.

The skin is made up of three main layers, the epidermis, dermis, and hypodermis. The sensation of touch is processed in the dermis, the middle layer located beneath the paper-thin outer layer, the epidermis. The dermis, also known as the true skin, is made up of nerve endings, sweat glands, oil glands, hair follicles, water, blood vessels, and protein. When any part of the skin touches something, the nerve endings located in the dermis carry messages to the brain about the object's texture, temperature, and feel. Some parts of the dermis are more sensitive than others.

11

SENSE DISORDERS

There are many disorders that can affect the body's ability to experience sensations. Sense disorders, although not life threatening, can diminish one's quality of life. The loss or impairment of any of the five senses can be a traumatic and depressing condition.

AGING CHANGES

Like other body systems, the senses undergo changes as a person ages. Hearing is the sense that is most commonly affected as a person becomes older. In fact, one out of every three Americans over the age of sixty and one out of two over the age of eighty-five have hearing loss.

The condition of gradually losing hearing as one ages is known as *presbycusis* (prez-bih-KYOO-siss). The condition seems to run in some families. Presbycusis occurs as the structures in the inner and middle ear suffer from wear and tear. Some people wear hearing aids in order to maintain their hearing.

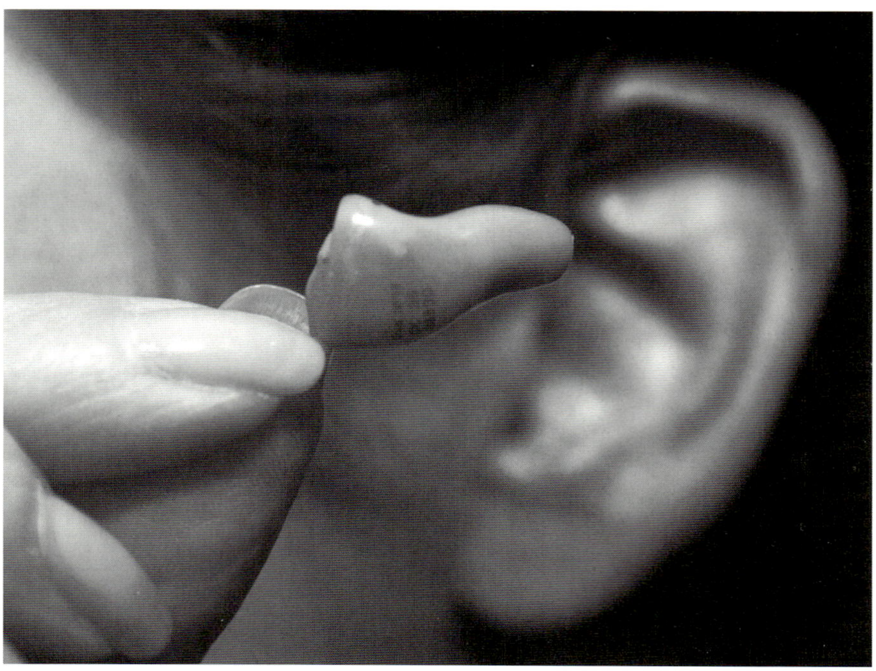

This is a type of in-the-canal hearing aid that is used by many people to reverse the effects of hearing loss.

Sight is also frequently affected as a person ages. Vision may change as parts of the lens become clouded, a condition known as a *cataract.* In addition, the pupils become less responsive to light and dark, and eye muscles may weaken.

Taste, smell, and touch are also affected by the aging process. As people grow older, their senses of taste and smell become less keen. The tongue loses working taste buds. The sense of touch also changes. Elderly people may become better able to tolerate pain but less able to tell the difference between cool and very cold things.

HEARING DISORDERS

There are a number of medical conditions that affect the ear. These conditions can cause hearing loss and bring on other medical problems. Hearing loss is a particularly common problem in the United States. According to the National Institutes of Health (NIH), about 28 million Americans currently suffer from some hearing loss.

One of the most common causes of hearing difficulties is a buildup of earwax in the ear canal. Excessive earwax buildup may require removal by a doctor. Other hearing and balance disorders include the following:

TYPE OF DEAFNESS	Nerve deafness
NATURE	Caused by damage to the nerves within the cochlea or the cochlear nerve, the nerve that passes messages to the brain; may affect one or both ears
TREATMENT	Condition is irreversible; cochlear implant may be possible
TYPE OF DEAFNESS	Sudden deafness
NATURE	Deafness occurs rapidly, either all at once or within three days; many different causes, including illness, trauma, and environmental causes
TREATMENT	Condition may heal by itself; doctors may prescribe steroids or, if the problem persists, hearing aids
TYPE OF DEAFNESS	Presbycusis
NATURE	Loss of hearing that occurs naturally as the body ages
TREATMENT	Condition is irreversible; hearing aids may help

Cochlear Implants

For some people with serious hearing loss or complete deafness, cochlear implants provide the opportunity to experience sound. A *cochlear implant* is very different from a hearing aid. A small device is implanted in the skin behind the ear. The device consists of a microphone to detect sounds; a speech-processor to select and sort sounds, a receiver and transmitter that receive sounds and turn them into electrical impulses, and electrodes that send the electrical impulses to the brain. The device does not restore hearing. Instead, it gives deaf and hearing-impaired people the chance to experience sounds and speech as interpreted by the device's speech-processor. Developed in the 1970s, the cochlear implant is now being used by thousands of Americans.

A cochlear implant is a device that allows seriously hearing-impaired people to hear sounds. Cochlear implants are powerful computers that analyze speech and other sounds and transmit them to a device implanted behind the ear, where the signal is relayed to the brain via connections to the person's auditory nerve.

DEAFNESS

Deafness is the total or partial loss of hearing in one or both ears. Deafness can result from several different causes, such as aging; hereditary factors; certain diseases such as rubella, meningitis, mumps, and damage to the ear. Types of deafness include the following:

EAR TUMORS

Tumors, both benign and malignant, can grow within the ear. Even when an ear tumor is benign, it can cause serious problems as it grows. Ear tumors within the inner ear, for example, put pressure on cells that affect hearing. This can cause hearing problems, including tinnitus (ringing in the ears) and loss of hearing in one or both ears. If a tumor grows too large, it may press on the brain, becoming potentially life threatening.

One type of benign tumor that can cause hearing problems is an *acoustic neuroma* (noor-OH-muh). This type of tumor grows on the nerve that carries messages from the ear to the brain. Acoustic neuromas, among the most common type of brain tumors, are a genetic disorder. Symptoms include tinnitus, vertigo, and hearing loss in the affected ear. Whenever possible, these tumors are surgically removed. However, any hearing that has been lost as a result of the tumor will not be restored by the surgery. For those who are unable to tolerate the surgery, such as elderly or very ill patients, doctors may use radiation therapy to shrink the tumor or slow its growth.

In rare cases, a benign tumor may become malignant. Symptoms of a tumor that has turned cancerous include rapid growth of the tumor, sudden pain in the ear, tingling or numbness of the ear, and hearing loss.

OTOSCLEROSIS

Otosclerosis (oh-toh-skluh-ROH-siss) is a condition in which a bony growth in the middle ear can result in deafness. The growth prevents the stapes from vibrating. Doctors are unsure about what causes otosclerosis, but the tendency to develop this condition may run in some families. This common condition affects one in every ten Americans.

Symptoms of otosclerosis include hearing loss, dizziness, and tinnitus. Doctors may treat the condition with surgery to bypass the growth, remove the stapes, and replace it with an artificial stapes that is not affected by the growth. A patient who is too elderly or too ill to tolerate surgery may need to wear a hearing aid that will help restore the ability to hear.

SIGHT DISORDERS

Blindness and impaired sight may be the result of illness or damage to the eyes or the parts of the brain that process sight messages. The most common causes of eye injuries are accidents in which the eye is cut by pieces of metal. Many of these injuries occur in the workplace. The most common types of sight disorders are the following:

TYPE OF DISORDER	Myopia (mye-OH-pee-uh), or nearsightedness
NATURE	Objects that are nearby are clearly visible, while those that are farther away are less clear
TREATMENT	Eyeglasses or contact lenses; corrective surgery
TYPE OF DISORDER	Hyperopia (hye-per-OH-pee-uh), or farsightedness
NATURE	Objects that are farther away are clearly visible, while those that are nearby are less clear
TREATMENT	Eyeglasses or contact lenses; corrective surgery
TYPE OF DISORDER	Astigmatism (uh-STIG-muh-tiz-um)
NATURE	Vision is blurred as a result of an irregularly shaped cornea
TREATMENT	Eyeglasses or contact lenses; corrective surgery
TYPE OF DISORDER	Glaucoma (glaw-KOH-muh)
NATURE	Aqueous humor does not drain properly, causing blockage and blindness
TREATMENT	Medications or surgery
TYPE OF DISORDER	Strabismus (struh-BIZ-muss)
NATURE	Weak or abnormal eye muscles result in eyes looking in different directions
TREATMENT	Eye exercises to strengthen muscles; corrective surgery
TYPE OF DISORDER	Amblyopia (am-blee-OH-pee-uh)
NATURE	One eye does not work as well as the other
TREATMENT	Eye exercises to strengthen muscles; corrective surgery

Surgical Procedures for Sight Disorders

In the past, people with vision problems had little choice but to wear glasses to correct the problems. Today, several surgical procedures can be performed, depending upon a person's disorder and physical condition.

- During a *corneal transplant,* doctors remove a patient's defective cornea, replacing it with a healthy one from a donor who has recently died. The procedure is usually very successful in restoring partial vision. Corneal transplants are the most common transplant procedures performed in the United States.
- *LASIK surgery* is a procedure in which eye doctors use a special laser to reshape the cornea by removing tissue. This type of surgery is commonly called refractive surgery, because it affects the focusing function of the eye. Not everyone is eligible for this procedure, but those who are often have nearly perfect vision after the operation.
- During most *cataract surgeries,* doctors remove the lens that has clouded over and replace it with an artificial one. A special type of cataract surgery, called phacoemulsification (fay-koh-ee-mul-sih-fih-KAY-shun), uses vibrations to break up a patient's cataracts, which are then suctioned from the eye.

SMELL DISORDERS

Smell disorders range from a mild, temporary loss of smell to a complete and permanent loss. A reduction in the ability to smell is called *hyposmia* (hye-PAHZ-mee-uh). A complete loss of the ability to smell is known as *anosmia* (an-AHZ-mee-uh). The loss of the sense of smell can be a serious, even life-threatening problem. Without this important sense, people are unable to detect poisonous gases or the scent of smoke from a fire. Other smell disorders include smelling odors that are not there and interpreting a normally pleasant smell as foul.

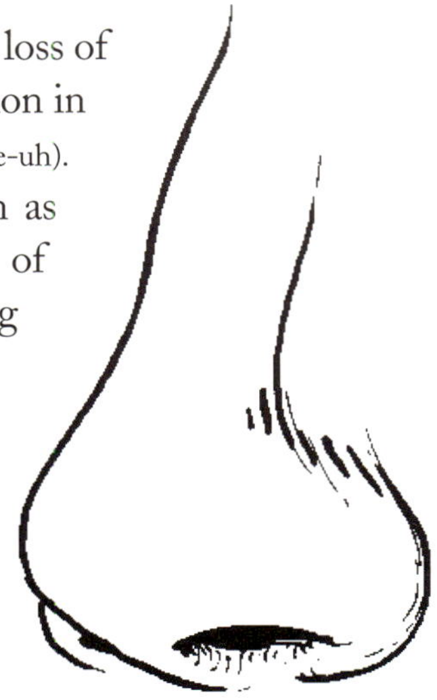

Some people are born with a poor sense of smell. For others, nasal congestion brought on by colds, flu, or allergies can cause a temporary loss of smell. The sense of smell usually returns unharmed after nasal congestion has been treated. Injuries (particularly head injuries), nasal polyps (PAHL-ipz), nasal tumors, nasal deformities, brain tumors, and Alzheimer's disease can all cause a person to lose the ability to smell. Radiation treatment, nasal or sinus surgery, and other medical procedures can also cause a person to lose the sense of smell. In some cases, doctors may never determine the cause.

Depending upon the cause of the disorder, people may or may not be able to recover their sense of smell. If the problem stems from a nasal defect, for example, treatment of the defect could solve the problem. In the case of a head injury, damaged olfactory cells may reform.

TASTE DISORDERS

Although the sense of taste might not seem as important as some of the other senses, taste disorders can be serious. A person without this sense is unable to use flavor clues to detect poison, spoiled foods, or foods to which they are allergic.

Several disorders can affect a person's sense of taste. Some people are born without the ability to taste. Others lose this sense as the result of injuries, which can affect the way that flavor messages are sent to the brain or the ability of the brain to interpret the messages. Such injuries may permanently impair a person's taste.

The most common cause of temporary taste disorders is medications. Problems can also be caused by certain medical conditions, including nasal congestion from colds and allergies, gum disease, and viral or bacterial infections. When the problem-causing medications are stopped or the underlying medical condition is treated, the sense of taste usually returns.

TOUCH DISORDERS

Some touch disorders involve losing sensitivity in an area of the body. People may lose their sense of touch as a result of an injury to the skin, nerves, or brain. Injuries can also result in numbness, tingling, or a "pins and needles" sensation in the affected body area. While these feelings may be the temporary result of repetitive movements, they may also indicate that something much more serious is wrong. Numbness and tingling, for example, are symptoms of heart attacks, strokes, and blood clots. People with these symptoms should seek medical attention immediately.

Other disorders cause the skin to become sensitive and irritated. Rashes, for example, cause uncomfortable itching. Rashes may be the result of any number of causes, including other medical conditions, medications, viral, bacterial, or fungal infections; and allergies.

One condition that can cause either numbness or extreme sensitivity of the skin is neuralgia. *Neuralgia* (noor-AL-juh) is a disorder in which a nerve to the area is affected. Neuralgia may be caused by a blood vessel or blood clot pressing on the nerve. In some cases, the nerve is damaged by chemicals or other medical conditions. Many times, however, doctors are unable to determine what is causing neuralgia. Elderly people are especially at risk of developing neuralgia. In some cases, neuralgia improves without treatment. In other cases, doctors recommend painkillers and medical, surgical, and therapeutic procedures.

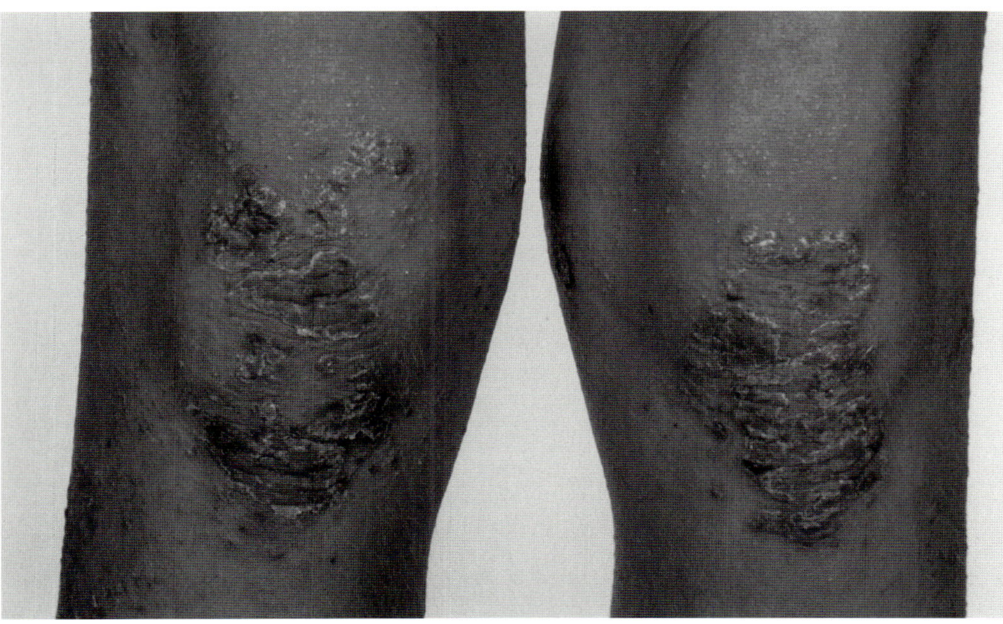

Rashes can be the result of a variety of causes, including reactions to medications, bacteria, fungal infections, and allergies.

DIAGNOSING SENSE DISORDERS

Many tests and devices are used to diagnose disorders of the senses and sensing organs. Some diagnostic tools are the following:

- Auditory brainstem response (ABR): This hearing test measures how well the brainstem and brain interpret sound.
- Scratch and sniff tests: Patients smell swatches containing certain odors to see which ones they can and cannot detect.
- Sip, spit, and rinse tests: Doctors sometimes check the ability to taste by applying chemicals directly to certain parts of the tongue.
- Tonometry (toh-NAHM-uh-tree): Tonometry is an eye test used to screen people for glaucoma. The test measures the pressure inside the eye. An increase in inner eye pressure may indicate glaucoma.
- Nerve conduction study/electromyography (ee-lek-troh-mye-AH-gruh-fee), NCS/EMG: This two-part test is used to diagnose nerve problems that may be causing pain and affecting the sense of touch. The first part of the test uses electrodes to test nerve activity. The second part of the test measures electrical activity in the affected area.

12

HOW THE ENVIRONMENT AFFECTS THE SENSES

Agents in the world around us can seriously affect the senses. Chemicals and pollution, cigarettes and alcohol—even loud noises—can all cause sensory damage.

CHEMICALS AND POLLUTION

Chemicals and pollution can seriously affect all of our senses. Exposure to certain chemicals (including some medicines), for example, can lead to a loss of taste. Lead poisoning, which is especially harmful to the brain and nervous system, can cause smell and hearing disorders. Lead can be found in the paint on the walls of older homes, as well as in older plumbing fixtures. It can also be found in contaminated dust and soil.

Mercury is an element that is found naturally in the environment. It can also be found in glass thermometers, batteries, and some paints. Exposure to large amounts of mercury can cause serious sensory and other problems. When pregnant women are exposed to mercury, their infants may be born deaf and have other serious medical problems.

Other agents that are known to cause sensory disorders include formaldehyde (form-AL-deh-hyde), carbon monoxide, arsenic, and tin. Specific disorders caused by chemicals and pollution include the following:

OTOTOXICITY

Ototoxicity (oh-toh-toks-IH-sih-tee) is poisoning of the inner ear that occurs when people come into contact with certain toxic chemicals or drugs. Depending upon the chemical, the condition can affect hearing, balance, or both. Chemicals that are known to cause ototoxicity include mercury, carbon monoxide, lead, manganese, and tin. Drugs that have been shown to cause ototoxicity include aspirin, quinine, certain antibiotics, and some anticancer medications.

Common symptoms of ototoxicity include ringing in the ears, dizziness, loss of balance, and hearing loss. Symptoms may be temporary, disappearing after a person stops taking the medicine that caused the problem. In other cases, ototoxicity causes permanent damage to a person's sense of hearing or balance.

OTITIS EXTERNA

Also called swimmer's ear, *otitis externa* (oh-TYE-tiss ex-TER-nuh) is an infection or irritation of the outer ear or ear canal. The condition is often caused by swimming in polluted waters. In other cases, it is the result of damage to the outer ear by foreign objects, such as cotton swabs or other items used to clean the ear. The infection, which is fairly common, most often affects teenagers and young adults.

Symptoms of otitis externa include ear pain, drainage of pus from the ear, and itching of the outer ear or ear canal. Doctors usually treat the condition with antibiotic eardrops to kill any infection-causing bacteria inside the ear. They may also use topical steroids to stop the ear from itching. If left untreated, the infection can spread to other parts of the body.

In spite of warnings about unsafe swimming conditions, a boy surfs in the Atlantic Ocean off Rio de Janeiro, Brazil, where a major sewer line is under repair. Sewage and other pollutants can spread disease to swimmers and others who come into contact with polluted ocean waters.

SMOKING AND THE SENSES

Tobacco products can impair people's ability to enjoy the world around them. Heavy smoking, for example, can cause irritations of the tongue and lead to a loss of taste. More serious health conditions of the mouth that are associated with the use of tobacco products include cancer of the mouth, most often of the lips or tongue. As many as eight out of ten cases of oral cancer are associated with tobacco use. Heavy alcohol use has also been linked to this type of cancer.

> **Fast Fact**
>
> Cataract surgery is the most common type of surgery in the United States.

Smoking can also affect a person's sight. Studies have shown that smoking increases a person's risk of developing cataracts later in life. Excessive alcohol use has also been linked to an increased risk of cataracts. Heavy smoking has also been linked to a reduction in the ability to smell.

Secondhand tobacco smoke can harm others, especially small children. Secondhand smoke can increase a child's risk of ear infections. In addition, secondhand smoke has been shown to raise the chances of children developing respiratory infections, SIDS, and lung cancer.

LOUD SOUNDS

Loud sounds can seriously damage hearing health. Ears can be hurt by a single, superloud noise, such as a gunshot at close range or a rock and roll concert. Some people are also exposed to loud noises over a series of years. For example, machine workers, firefighters, and police officers are all exposed to loud noises in the workplace.

TINNITUS AND NOISE-INDUCED HEARING LOSS

Two conditions that are caused by loud sounds are tinnitus and noise-induced hearing loss. *Tinnitus* (TIN-ih-tuss) is a condition that causes a person to constantly hear ringing, roaring, clicking, or hissing noises in the ear. People with the condition may have difficulty hearing, sleeping, or working. Tinnitus is often caused by exposure to loud noises, such as close-range gunshots or very loud music. It may also be caused by hearing loss, some medicines, and other health problems.

Noise-induced hearing loss (NIHL) is temporary or permanent hearing loss that results from consistent exposure to loud noises over a period of time. (Hearing loss caused by one-time exposure to a loud noise is called *acoustic trauma.*) The hearing loss results from damage to the structures of the ear.

Unfortunately, both tinnitus and NIHL are common conditions in the United States today. According to the National Institute on Deafness and Other Communication Disorders (NIDCD), tinnitus currently affects at least 12 million Americans. There is no cure for either tinnitus or NIHL. Doctors may prescribe a hearing aid to improve the hearing. For tinnitus, a device called a masker may also be used. Placed behind the ear, a masker covers some of the noises associated with tinnitus. Additionally, some medicines ease the effects of tinnitus for some patients.

Charlie Haden, leader of the jazz group Quartet West, suffers from the hearing problems tinnitus (ringing in the ears) and hyperacusis (supersensitivity to loud sounds). He plays concerts with a Plexiglas shield between himself and the drums, piano, and saxophone.

Hear, Hear

Now hear this: Loud music can damage your hearing. Just ask Lars Ulrich, drummer of the band Metallica. After years of playing loud music at concerts around the world, Ulrich now suffers from tinnitus. He's not alone. As many as 30 percent of all rock and rollers suffer from some type of hearing loss.

In 1988, California rocker Kathy Peck helped found a group called Hearing Education and Awareness for Rockers (HEAR). Like Ulrich, Peck suffered hearing loss after years of playing loud music. Peck's group provides education and information about NIHL. HEAR hopes to let kids know that the sense of hearing is a precious gift that needs to be protected.

VIRUSES, BACTERIA, AND FUNGI

Viruses, bacteria, and fungi can seriously affect the body's senses. Infections can cause hearing, balance, taste, and vision losses.

OTITIS MEDIA

Otitis media, more commonly known as ear infection, occurs when viruses or bacteria infect the middle ear, causing swelling and irritation. Fluid and pus become trapped inside the ear, resulting in pain and discomfort. Otitis media is a common condition, especially among children. In fact, it is the most common illness among babies and young children.

Symptoms of otitis media include ear pain, fever, hearing problems, nausea, vomiting, and diarrhea. Otitis media can also lead to an infection of the eustachian tubes and the adenoids. In serious cases, excess fluid in the middle ear can cause the eardrum to rupture, leading to severe pain or drainage of fluid or pus from the ear. A complication of chronic, or recurring, otitis media is the development of a cyst in the middle ear. If the cyst is not surgically removed, it may become infected. Left untreated, the cyst may erode the bones of the middle ear.

Treatment for otitis media includes antibiotic medication to kill the bacteria. In cases of viral infection, doctors may prescribe painkillers and other measures to ease the pain. Some children may require surgery to correct the problem. Surgery is often recommended when a child develops the disease a number of times. The most common type of surgery for the condition is called a *myringotomy* (mir-ing-AH-tuh-mee). During the procedure, doctors make a small cut in the eardrum and insert tubes into the child's ear. The tubes, which prevent pressure from building up within the ear, eventually fall out on their own. In other cases, doctors perform surgery to remove a child's adenoids. The adenoids, when infected, can block the eustachian tubes, causing ear problems.

Fast Fact

Nearly 85 percent of all children in the United States will experience otitis media by the age of three.

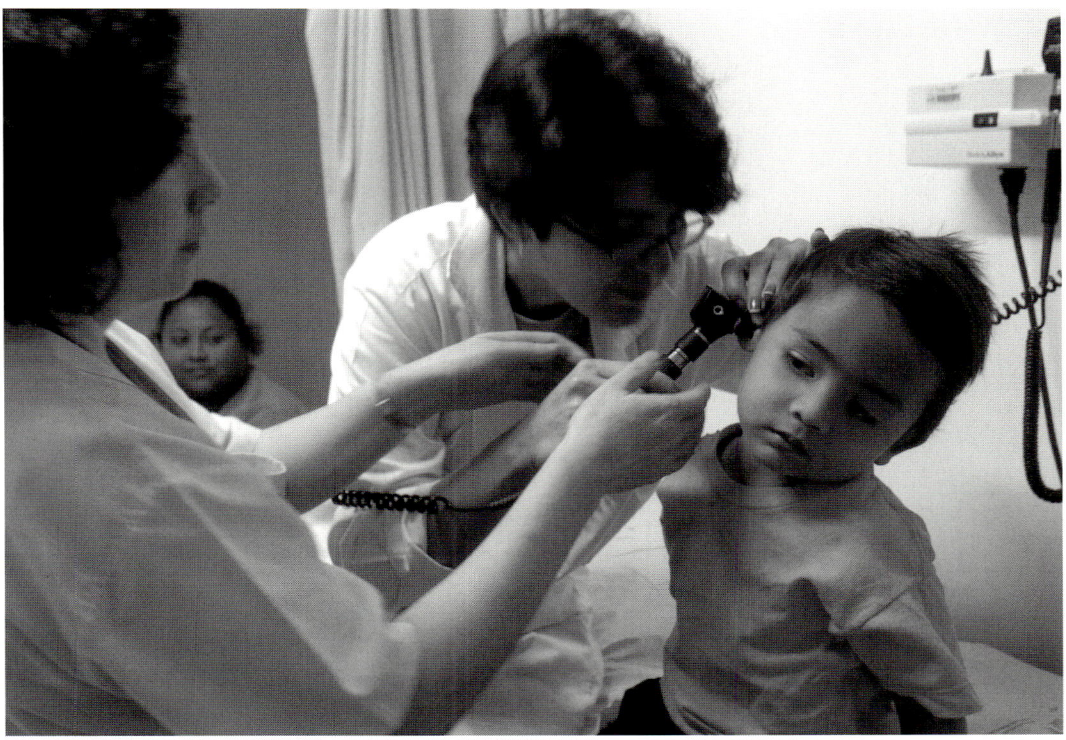

Two doctors examine this five-year-old who is suffering from otitis media, also known as an ear infection. Early treatment of ear infections prevents them from becoming more serious.

CONJUNCTIVITIS

Conjunctivitis (kun-junk-tih-VYE-tiss) is an infection of the *conjunctiva,* the membrane that lines the inner eyelid. The infection is most often caused by viruses but may also be caused by bacteria and fungi. Allergies, tobacco smoke, and dust may also cause conjunctivitis. Viral conjunctivitis is also known as pinkeye.

When caused by viral or bacterial infection, conjunctivitis is very contagious. For this reason, people with the condition should avoid rubbing or touching their eyes, wash their hands frequently, and avoid sharing towels, handkerchiefs, or cosmetics with others. Symptoms of conjunctivitis include a reddened eye, discharge from the eye, pain, itching, or a gritty feeling in the eye, and blurred vision. Doctors often treat bacterial conjunctivitis with antibiotics that can be placed in the eye to kill the bacteria. Other treatments include warm compresses and eyedrops to relieve pain and irritation.

> **Fast Fact**
>
> Neonatal conjunctivitis occurs when a baby's eyes are infected with bacteria during labor. Children born with this condition must be treated immediately to prevent permanent blindness.

STIES

Sties are red, painful infections of the oil glands of the eyelids. These infections are most often caused by *Staphylococcus* (staff-ih-loh-KAHK-uss) bacteria on the skin around the eye. Symptoms include swelling, sensitivity to light, pain, and excess tearing. While sties often disappear without treatment, warm compresses and antibiotic creams can speed the healing process.

13

KEEPING THE BODY HEALTHY

Your body is your most valuable asset. It's important to take care of it and keep it working as well as possible. For most people, caring for the body is not difficult to do. Even the simplest steps, taken now, can prevent problems later in life. For example, washing your hands to prevent the spread of germs, wearing a seatbelt to minimize the risk of traumatic injuries, and choosing not to use tobacco products will help keep the body safe and healthy. Following a healthy diet, getting plenty of exercise, and making good lifestyle choices will also help keep your body in top form.

DIET

You can keep your body healthy by eating a nutritional diet. Some nutrients are especially important in keeping the endocrine, reproductive, and nervous systems healthy. Calcium-rich diets are important for women experiencing menopause, for example. During menopause, the body produces less estrogen, which helps the body use calcium. Another important nutrient is iodine, essential to a healthy endocrine system. People with a lack of iodine in the diet can suffer goiters and other thyroid-gland problems.

> **Fast Fact**
>
> Researchers say that eating a diet that is low in fat may prevent nervous system disorders that cause dementia.

Healthy diets include foods that are low in fat and sodium and rich in natural vitamins and iron. By eating a sensible diet, people can prevent obesity, a condition that can lead to an increased risk of diabetes.

Pregnant women must be especially careful to eat healthy, nutritious foods. They are nurturing their fetuses by providing them with the vitamins and minerals the fetuses need to develop properly. Doctors have linked good prenatal care, including a healthy diet and proper medical attention, to a decreased risk of cerebral palsy and other congenital conditions.

LIFESTYLE CHOICES

Alcohol and other drugs can take a serious toll on every part of the body. These drugs are especially dangerous to developing fetuses. Everything that a pregnant woman takes into her body is passed through the umbilical cord into the fetus. The effects of toxic substances on the growing fetus can be long-lasting and deadly. Infants exposed to drugs in the womb are more likely to be born early and have long-term mental and physical problems.

Self-testing is important to the health of the reproductive system. Men are encouraged to perform testicular self-exams to detect testicular cancer and other problems in that area. Women are encouraged to perform breast self-examinations to check for lumps, signs of breast cancer, and other conditions. Women over the age of forty are encouraged to get a mammogram once every year or two to aid in the early detection of breast cancer.

One important way to keep yourself healthy is to avoid too much stress whenever possible. A little stress is okay and can even be positive, but chronic stress is believed to raise the risk of developing certain medical conditions and to worsen others. Some medical conditions that may be caused or worsened by stress are diabetes, MS, Parkinson's disease, tension headaches, reproductive problems, endocrine disorders, and depression. Stress has also been linked to heart disease and some types of cancer.

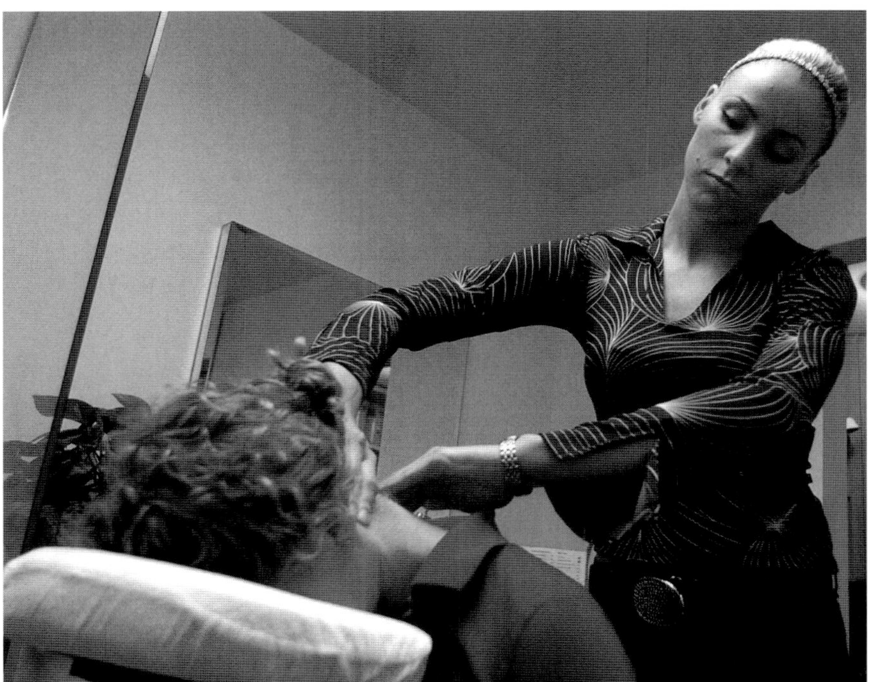

A massage therapist in Deerfield, Illinois, works on a fifteen-year-old high school client. Spa owners say that more and more teens are seeking massage and other ways to relieve stress.

For sexually active people, safe sex is important. This includes using condoms when engaging in intercourse. Latex condoms must be used properly to protect people from disease. Many health experts, including the U.S. Department of Health and Human Services, recommend abstinence, or not having sexual intercourse, as the best and only sure way to avoid such health problems as HIV infection, STDs, and hepatitis.

Exercising the Brain

Everyone knows that exercising your body is good for you, but did you know that it is also important to exercise your brain? Doctors say that as people age, the thinking process may slow down. Using the brain by reading or doing crossword puzzles and other activities that make you think twice can help you stay sharp.

Tips for Reducing Stress

Here are some tips from the experts on loosening up and lessening stress:

- Take deep regular breaths when you begin to feel anxious or stressed out.
- Talk about your feelings with people that you trust. Let them know what's bothering you.
- Follow healthy eating and sleeping habits whenever possible.
- Exercise regularly.

GLOSSARY

acquired immune deficiency syndrome (AIDS)—a serious, life-threatening condition; people with AIDS suffer from a weakened and damaged immune system, which makes them susceptible to many other diseases

autoimmune disorder—a condition caused when the body's immune system attacks other parts of the body that it mistakenly identifies as foreign or harmful

bacteria—microscopic, single-celled organisms that are found everywhere; many bacteria cause disease, but not all are harmful

benign—not cancerous

biopsy—a surgical procedure to remove tissue, cells, or fluids from the body for examination

cancer—a disease in which a group of abnormal cells in the body divide without control

chemotherapy—the use of anticancer drugs to cure or stop the spread of cancers

computerized tomography (CT)—a test that uses a thin beam of electromagnetic radiation to create a three-dimensional picture of what's inside the body

endocrine disruptors—chemicals, whether synthetic or naturally occurring, that interfere with the proper functioning of the endocrine system

endocrine system—the body system that manufactures and releases hormones

environmental health—the body's reaction to conditions that people have little or no control over, including sunlight, bacteria, chemicals, and pollution; it may also include medical problems that result from personal lifestyle choices, such as smoking

fungi—parasitic organisms, such as mushrooms and molds, that feed on both living and decaying matter

genetic disorder—an abnormality in a person's genes or chromosomes

hormones—chemical messengers that regulate growth, metabolism, and other body functions

hormone-replacement therapy—a treatment in which patients are given artificial hormones to replace those that they are lacking

human immunodeficiency-virus (HIV)—a contagious virus that attacks the body's immune system; HIV can lead to AIDS

magnetic resonance imaging (MRI)—a test that uses magnets and radio waves to create an image of the inside of the body

malignant—cancerous or harmful

menopause —the period in a woman's life when hormone production slows down and she ceases having a monthly cycle

nervous system—the body system that carries, translates, and acts upon messages from around the body; made up of the brain, spinal cord, and a network of nerve cells called neurons

puberty—physical changes that boys and girls experience as they reach sexual maturity

radiation—energy in the form of rays or particles

reproductive system—the body system that is responsible for creating new life

senses—abilities that people (and other animals) have to fully experience the world around them; the body's five senses are hearing, sight, smell, taste, and touch; to make each sense possible, special organs work together with special nerve cells and the brain to interpret messages from the outside world

sexually transmitted diseases (STDs)—diseases that are passed from one person to another through unprotected sexual intercourse, whether oral, vaginal, or anal

toxin—poison

tumor—an abnormal mass of tissue that grows in the body; tumors may be benign or malignant

virus—a simple germ organism that is always harmful to people; viruses are found everywhere

X-ray—a test that uses electromagnetic radiation to take a two-dimensional picture of what's inside the body

BIBLIOGRAPHY

BOOKS

Avraham, Regina. *The Reproductive System.* Philadelphia: Chelsea House, 2000.

Brynie, Faith Hickman. *Perception.* Woodbridge, CT: Blackbirch Press, 2001.

Clayman, Charles, ed. *The American Medical Association Encyclopedia of Medicine.* New York: Random House, 1989.

———. *The Human Body: An Illustrated Guide to Its Structure, Function, and Disorders.* New York: DK Publishing, 1995.

Cothran, Helen. *Teen Pregnancy and Parenting.* San Diego: Greenhaven Press, 2001.

Kittredge, Mary. *The Human Body: An Overview.* Philadelphia: Chelsea House, 2001.

Little, Marjorie. *The Endocrine System.* Philadelphia: Chelsea House, 2000.

Nolte, John. *The Human Brain in Pictures and Diagrams.* St. Louis: Mosby, 2000.

Silverstein, Alvin, Virginia Silverstein, and Laura Silverstein Nunn. *Senses and Sensors* series. New York: Twenty-First Century Books, 2001.

Walker, Pam, and Elaine Wood. *The Brain and Nervous System.* San Diego: Lucent Books, 2002.

Walker, Richard, ed. *Encyclopedia of the Human Body.* New York: DK Publishing, 2002.

WEB SITES

Alzheimer's Association
www.alz.org/

American Diabetes Association
www.diabetes.org/main/application/commercewf

CDC National Center for Environmental Health (NCEH)
www.cdc.gov/nceh

The Children's Environmental Health Network
www.cehn.org/cehn

The Endocrine Society
www.endo-society.org/

KidsHealth
kidshealth.org

NIH National Eye Institute (NEI)
www.nei.nih.gov/

NIH National Institute of Child Health and Human Development (NICHD)
www.nichd.nih.gov/

NIH National Institute of Neurological Disorders and Stroke (NINDS)
www.ninds.nih.gov/

NIH National Institute on Deafness and Other Communication Disorders (NIDCD)
www.nidcd.nih.gov/

NIH National Toxicology Program Center for Evaluation of Risks to Human Reproduction (CERHR)
cerhr.niehs.nih.gov/

United States National Library of Medicine
medlineplus.nlm.nih.gov

INDEX

Note: Page numbers in *italics* refer to maps, illustrations, or photographs.